A CHILD
IS BORN

A CHILD IS BORN

Photography by Lennart Nilsson and Linda Forsell

Text by Lars Hamberger and Gudrun Abascal

Translated from the Swedish by Linda Schenck and Robin Blanton

A Merloyd Lawrence Book/Bantam
New York

CONTENTS

A CLASSIC LIVES ON 7

WOMAN AND MAN 11

The chemistry of love 12 • The human code 15 • X or Y? 19
The expectant mother 20 • The menstrual cycle 24 • Ovulation 27
The expectant father 32 • Inside the sperm factory 34

FERTILIZATION AND CONCEPTION 41

Journey toward the egg 42 • Exposing the egg 46 • A new human being is created 52
Journey through the Fallopian tube 55 • A soft landing in the uterus 61 • When nature needs help 65

PREGNANCY 73

Early signs 74 • Rapid, daily changes 76 • The early brain 78 • The first heartbeats 81
Confirmed! 85 • The body and face take shape 86 • Signals to and from the spine 91
A little human begins to take shape 92 • Eyes and ears 95 • Circulatory systems meet 96
The embryo becomes a fetus 102 • Comfortable or miserable? 104 • Factors that can affect pregnancy 108 • Prenatal care 113 • Growth spurt 118 • Everything all right in there? 123
Prenatal tests 126 • An aqueous life 129 • Halfway there 132 • Life in the womb 138
Almost ready for life outside 143 • Born early 144 • Getting to be a heavy load 150
Only a few weeks left 156 • Delivery time approaches 159 • Overdue 165

LABOR AND DELIVERY 169

First signs 171 • At the hospital 174 • Labor pains 182 • Methods of pain relief 184
Baby on the way! 188 • Welcome to the world 192 • C-sectioning into life 197
Aren't you lovely! 201 • The first days 206

LENNART NILSSON 211

INDEX 217

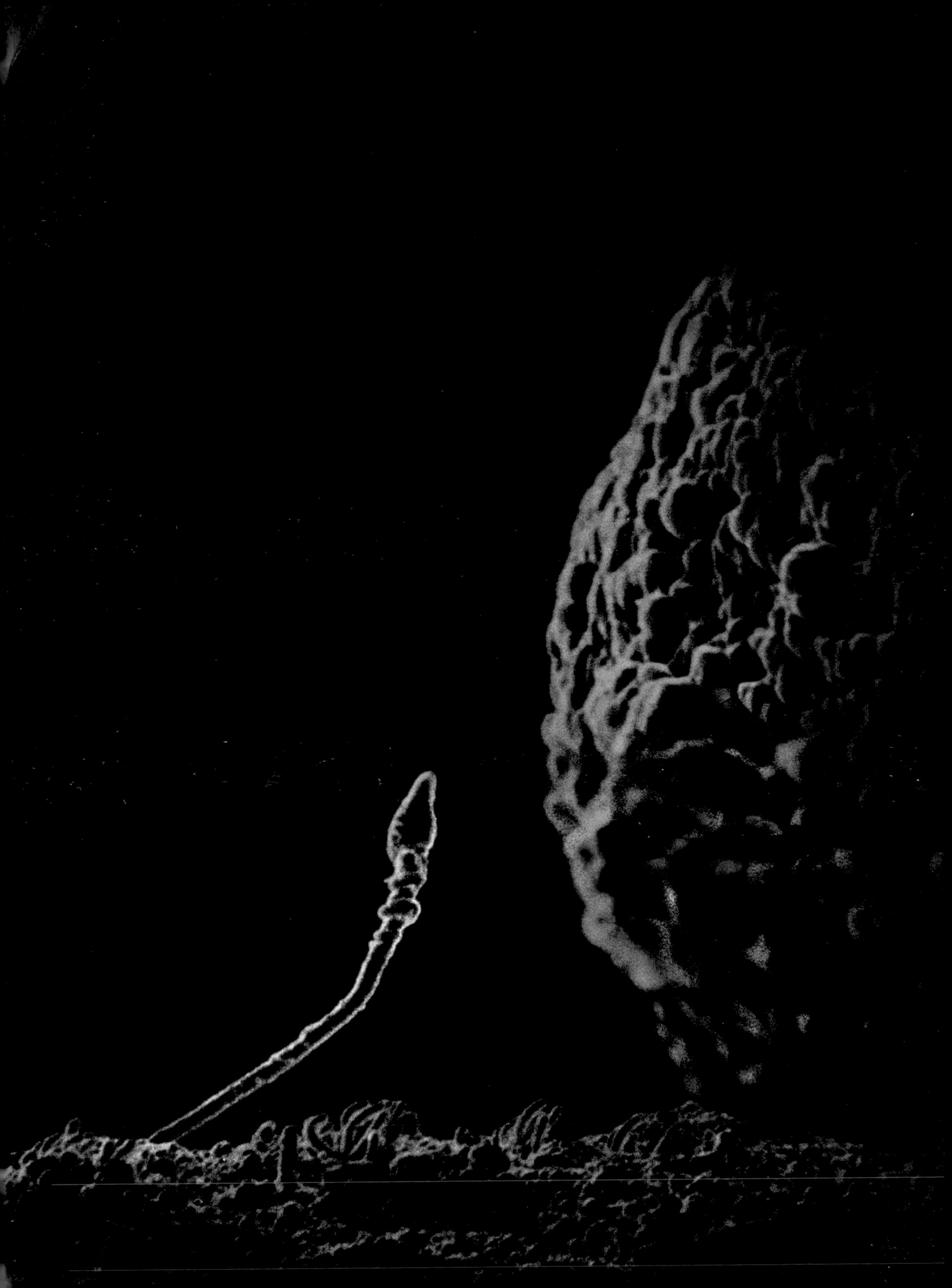

A classic lives on

The first edition of *A Child Is Born* was published in Sweden by the publishing house Albert Bonniers Förlag in October 1965, with epoch-making photographs by Lennart Nilsson and text by professor Axel Ingelman-Sundberg and associate professor Claes Wirsén. Readers followed the daily lives of parents-to-be over the course of a pregnancy. They watched an embryo grow in the womb, entered the delivery room alongside the parents, and saw the newborn experience its first days in the outside world.

Earlier the same year, *Life* magazine had published some of the photographs taken by Nilsson in an essay titled "Drama of Life Before Birth." It proved a milestone in the history of the magazine. The entire print run sold out in days. Soon the world would also see the first pictures of the earth taken from the moon. Like those iconic images, Nilsson's photographs forever changed the way we look at ourselves.

Over the years, *A Child Is Born* has been revised and published in several new editions. Nilsson took advantage of advances in technology and medicine to create new and better photographs and improved versions of his originals. Changes in maternity care and our growing understanding of the life of the embryo and fetus in utero were reflected in new texts written by new authors. The story was also filmed for television, reaching millions around the globe.

This is the fifth edition of *A Child Is Born.* It teams Nilsson's classic medical photographs with new documentary photographs by Linda Forsell and text by professor Lars Hamberger and midwife Gudrun Abascal. We began work on this book in 2016, the year before Lennart Nilsson's death. We believe he would have been proud to see the results.

Anne Fjellström
Lennart Nilsson Photography

Per Wivall
Senior editor, Bonnier Fakta

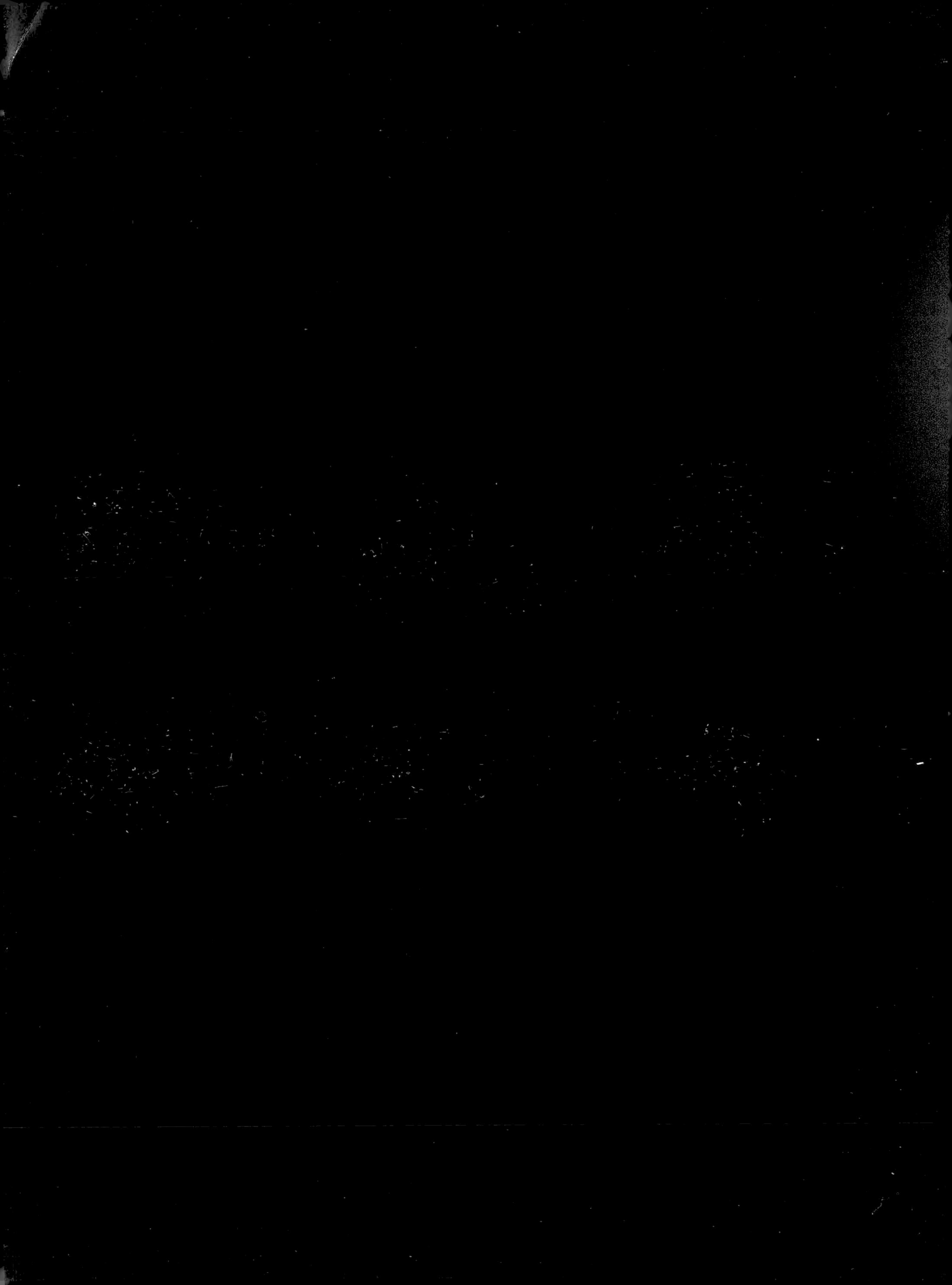

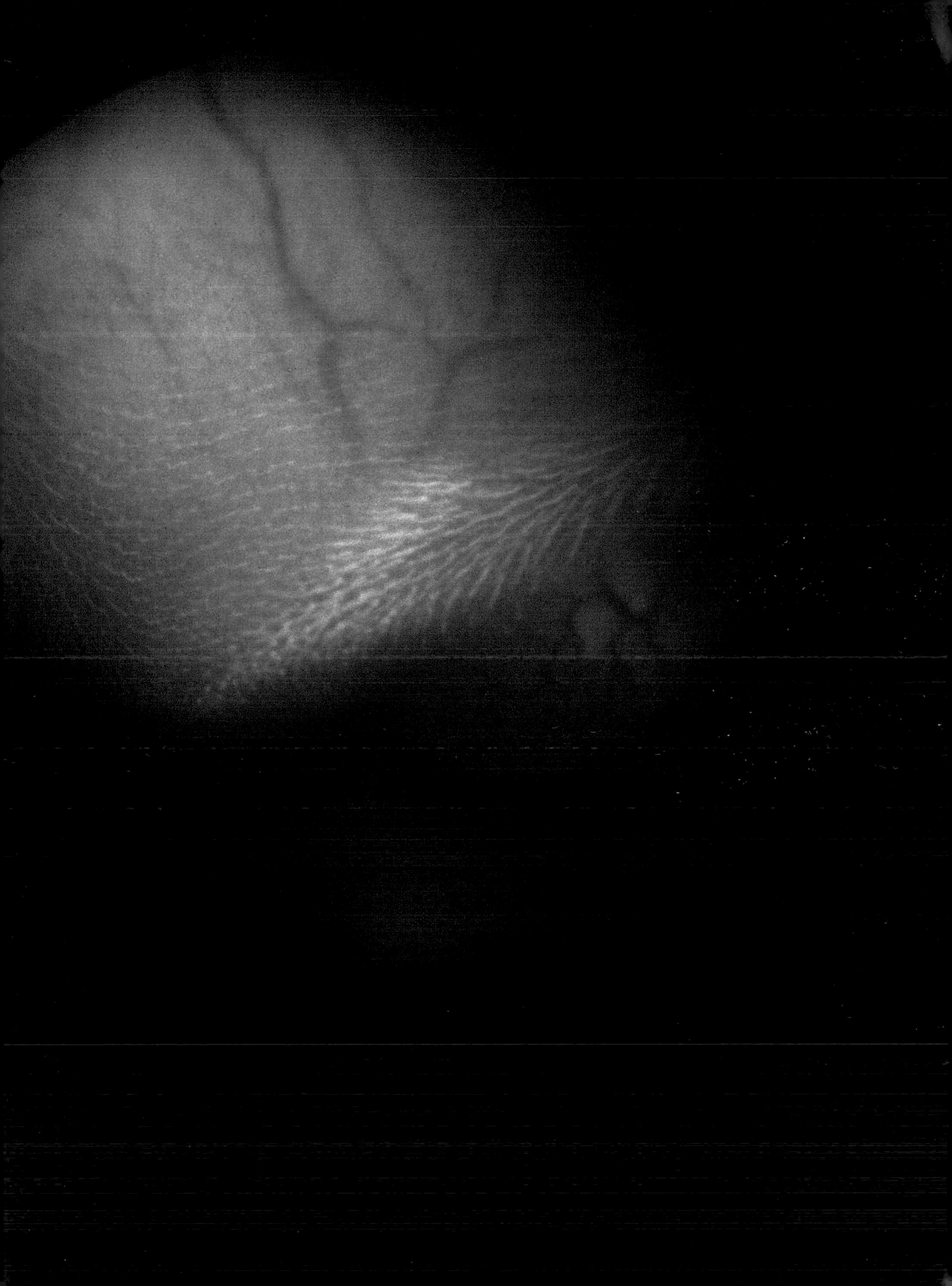

WOMAN AND MAN

Love—desire—longing. Sexual desire at its simplest is a biological drive, designed to ensure reproduction and the survival of our species, ideally with the best possible gene pool. Desire can be primitive, but also complex—as when two people fall in love. Love and sex bring us closer together, in body as well as spirit, instilling a feeling of deep connection.

Testosterone, the most important male sex hormone.

The chemistry of love

It all begins in the brain, with that tiny, magical spark of first attraction. Something about another person tugs at us: their looks, their charisma, the sound of their voice, the way they carry themselves, or the way they smell. It could be a glint in the eye, a lingering glance, or an infectious laugh. Or maybe it is just that the time is right, for us or for our biological clock. Whatever the cause, our attraction has real physical effects. We blush and stammer. Our palms sweat. We feel a tingling in our veins, butterflies in our stomach.

When it comes to attraction and our choice of partner, our biology and chemistry probably affect us more than we know. The chemistry of love involves a host of substances: dopamine, noradrenaline, serotonin, oxytocin, vasopressin, cortisol, pheromones, and especially the sex hormones estrogen and testosterone. Testosterone in men and estrogen in women transmit complicated chemical messages. These hormones affect our appearance and feelings and are essential to the reproduction process.

Like sexual desire, the longing for a child is a powerful, primal drive in women and men alike. In nature, almost all reproductive processes rely on one male and one female parent. This is true in both the plant and the animal kingdoms, which means it is true for humans as well. Conceiving a human baby requires a mature and viable egg from a woman and a mature and viable sperm from a man, although this may or may not be reflected in the actual partner relationship or the makeup of the family-to-be. For many couples, a child is the natural consequence of what began with attraction, then became passion, and finally grew into love.

Testosterone is produced in a man's testicles and governs, among other things, his sex drive.

> The female sex hormone estrogen has a great influence on a woman's body throughout the different stages of her life.

The human code

All human beings belong to the biological species *Homo sapiens*, with a shared genetic code that distinguishes us, for example, from apes, from pigs, and from birds. The great apes, chimpanzees, gorillas, and orangutans are our closest relations; our genetic codes differ only marginally from theirs. We are also genetically very similar to swine, or the porcine family. The differences from one human being to the next are even smaller, measurable in tenths of a percent, but they are still big enough to make each individual unique. We now know that even identical twins carry tiny genetic differences. Human beings of the same descent are genetically very similar; the closest likeness is among members of the same nuclear or extended family, who can share the same hair and eye color, height and weight, and even health status and life expectancy.

In recent years genetics has become a central focus in biology. Much can be explained by genetics, but it is important to realize that genetics and the environment are in constant interaction. We know that our environment affects us in many ways, and that the environment we experienced at the embryonic and fetal stages of our development influences our later lives. We are now very much aware of how important it is for pregnant women to think about what they eat and drink and the environments to which they are exposed.

Today we know much more than ever before about the significance of genetics, both for mankind as a unique species and for each individual. Our complete genetic code, consisting of some 20,000 different genes, has now been mapped. But we still know relatively little about what these genes do, what they imply for each of us, and how they either cooperate with or oppose one another in our bodies. And much remains to be learned about how the environment affects the ability of each individual gene to express itself.

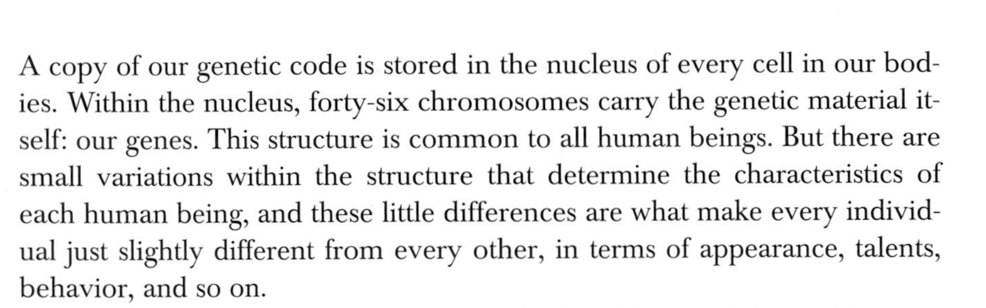

A copy of our genetic code is stored in the nucleus of every cell in our bodies. Within the nucleus, forty-six chromosomes carry the genetic material itself: our genes. This structure is common to all human beings. But there are small variations within the structure that determine the characteristics of each human being, and these little differences are what make every individual just slightly different from every other, in terms of appearance, talents, behavior, and so on.

Since the genetic material in every cell of each human being is identical, the details of that individual's genetic composition may be determined by examining any single cell. These techniques are used today to trace hereditary disposition for certain diseases.

Genetic material consists of DNA (deoxyribonucleic acid) molecules in the shape of a long double helix. These intertwined spirals of chemical building blocks are often designated by the letters A, C, G, and T. Combinations of these letters can spell out a very large number of different messages.

Our cells multiply by division, and each time a cell divides, two new ones, with exactly the same genetic material, are created. In every second in time, in every part of our bodies, throughout our entire lives, thousands of new cells, identical to the old ones, are being created. Once the cells of a bodily organ age, they expire, in accordance with a special pattern (known as programmed cell death, or apoptosis), and are replaced by new ones, keeping our bodies young and strong for many years.

If the human DNA chain were laid out flat, it would be 1.8 meters (2.3 yards) long. This chain, containing an astonishing million pieces of encoded information, is somehow enclosed in every nucleus of every cell in our bodies!

> Human chromosomes paired up at the instant of cell division. Each chromosome has two arms connected in the middle by a centromere. Each chromosome also has a specific appearance.

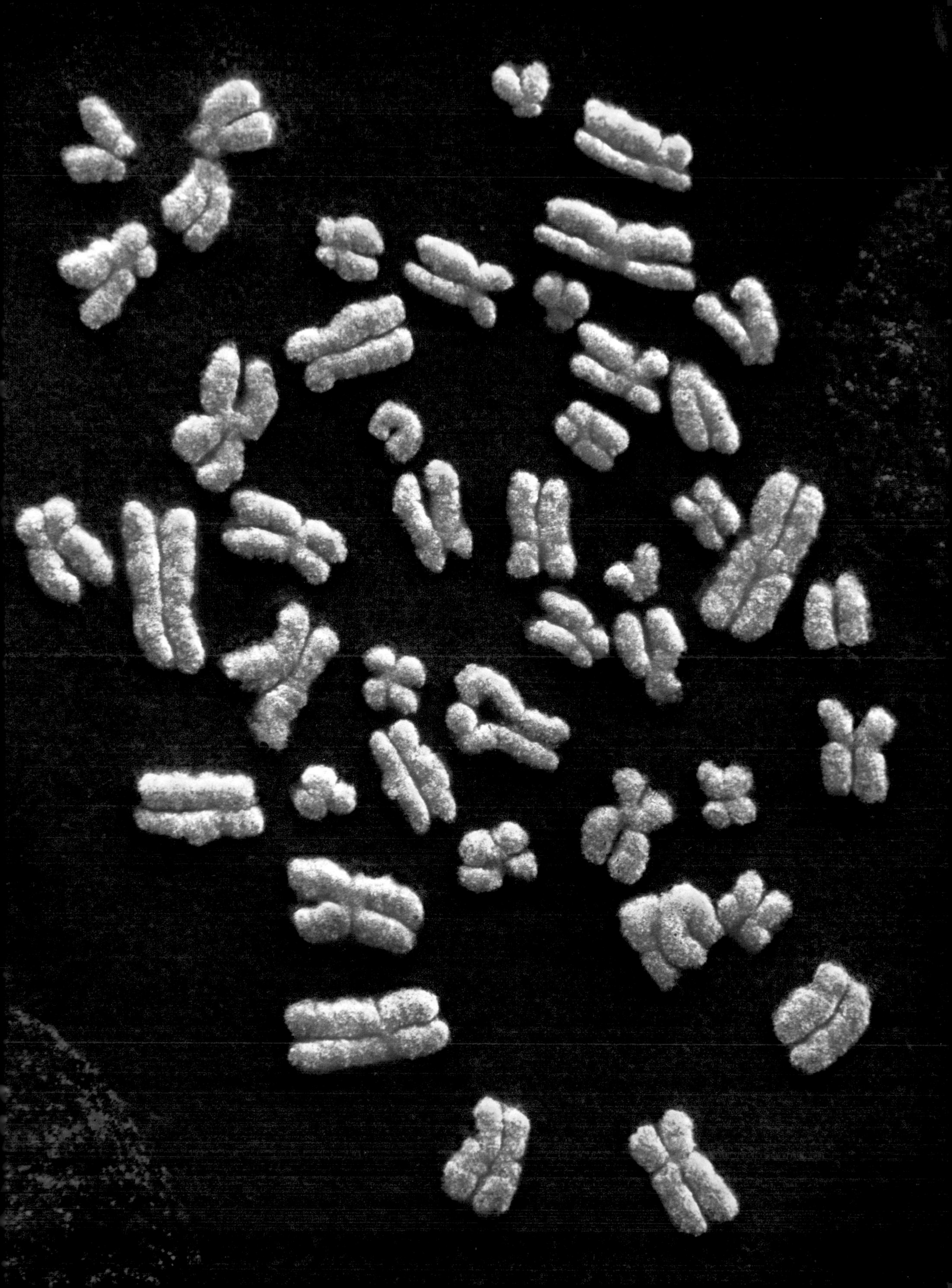

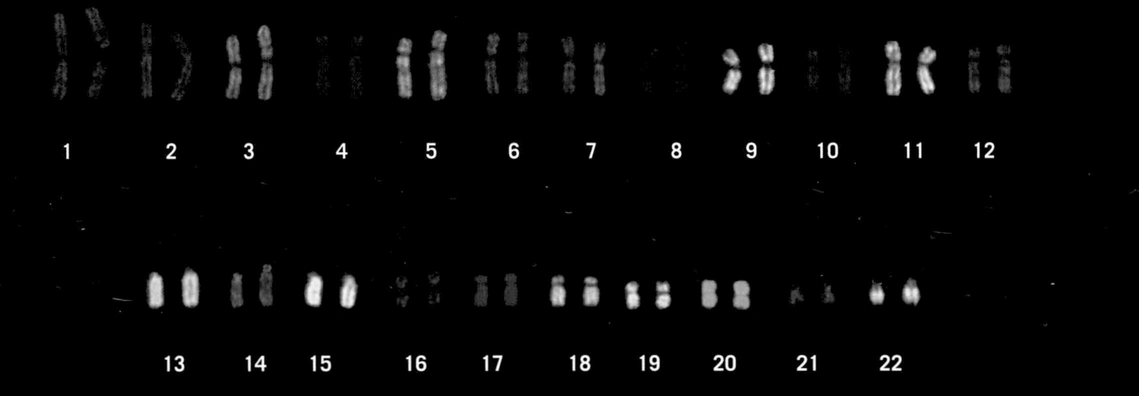

X X X Y

23

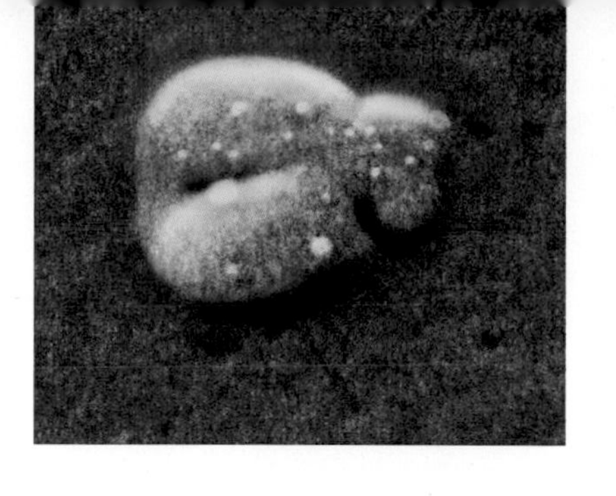

Y chromosome.

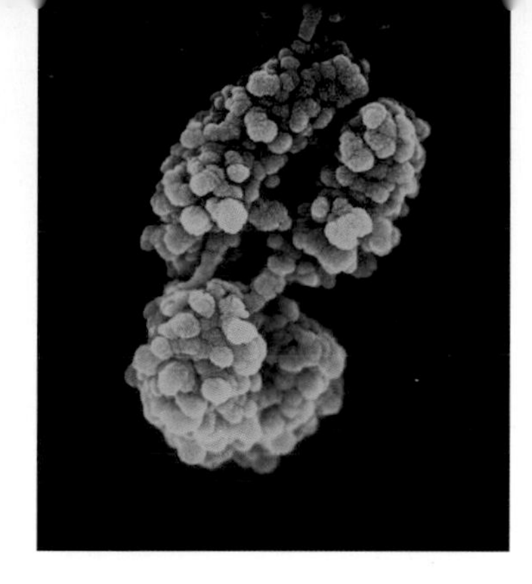

X chromosome.

X or Y?

The Y chromosome is the smallest human chromosome. It contains only about a hundred genes. The X chromosome is much larger.

< The chromosome pairs arranged in rows and displayed in different colors (chromosome painting). They are numbered from largest to smallest. Pair 23, the sex chromosomes, consists of either two X chromosomes (in a woman) or one X and one Y (in a man).

Sex cells differ from the other cells in the human body in that at the moment of fertilization they contain only twenty-three chromosomes each. When egg and sperm fuse, the same number of chromosomes come from the man and the woman, bringing the total back up to forty-six chromosomes, in twenty-three pairs. The first twenty-two pairs of chromosomes are the same in both sexes and they are numbered by size: chromosome 1 is the largest and chromosome 22 is the smallest. But the twenty-third pair is unique, because it can consist either of two X chromosomes in women, or one X and one Y chromosome in men.

Immature eggs, like all cells in the body, contain forty-six chromosomes, and the twenty-third pair is always two X chromosomes. But a few hours before ovulation, when the egg is almost ready to be fertilized, the number of chromosomes is reduced by half. Immature sperm also contain forty-six chromosomes, and this number is also halved when the sperm matures. But here something unique occurs, because the twenty-third chromosome pair in a man consists of one X and one Y chromosome. When the immature sperm splits into two parts, one contains an X chromosome and the other a Y. Thus half the mature sperm contain the genetic traits for girls and the other half for boys. The trait of the sperm that fertilizes the egg determines the sex of the new human being.

The expectant mother

After both the man and the woman have passed their own genetic material on to the new individual, the woman plays the main part until birth. In her body the first cell divisions take place, and a new little being is formed. Her reproductive system, which prepares for fertilization and pregnancy every month for about thirty-five years, is optimally adapted to nurturing an embryo and carrying it to term.

In contrast to a man, who produces sex cells throughout his life, a woman's stock of eggs develops before she is born, then decreases gradually. In the fourth month of embryonic development the ovaries of the female fetus have already produced the six to seven million eggs that comprise her lifetime production of eggs. Even before she is born, millions of the eggs expire; this programmed cell death in the ovaries continues steadily after birth. By puberty, when the young woman ovulates for the first time, only about 400,000 immature eggs remain, and by the time she reaches menopause, in her fifties, virtually her entire stock has been depleted.

During the entire period in which a woman is fertile and ovulating, her ovaries use up no more than four hundred eggs; this figure is often lower since pregnancies, nursing, and hormonal methods of birth control all prevent a woman from ovulating for certain periods. Only a few of these eggs at most will ever be fertilized and become a child. So a woman actually has an enormous reserve capacity.

Before puberty a girl's hormones undergo a change that makes her ovaries begin to produce more estrogen, and the eggs begin to mature. The shape of her girlish body also begins to change, and her entrance into adulthood is confirmed when she begins to menstruate. The most important female hormone is estrogen. It contributes not only to the development of a woman's figure but also the size of her breasts, the softness of her skin, and the thickness of her pubic hair; it also affects some regions of her brain. Estrogen reaches all these parts of her body via the circulatory system.

The egg, the female counterpart to the male sperm, is one tenth of a millimeter (0.0039 inch) in diameter, almost large enough to be visible to the naked eye. The ovum is a giant in the cellular microworld, but its nucleus is relatively tiny—no larger than the head of a sperm, or about the same size as other cells in the human body. The nucleus is where the woman's genetic material is stored.

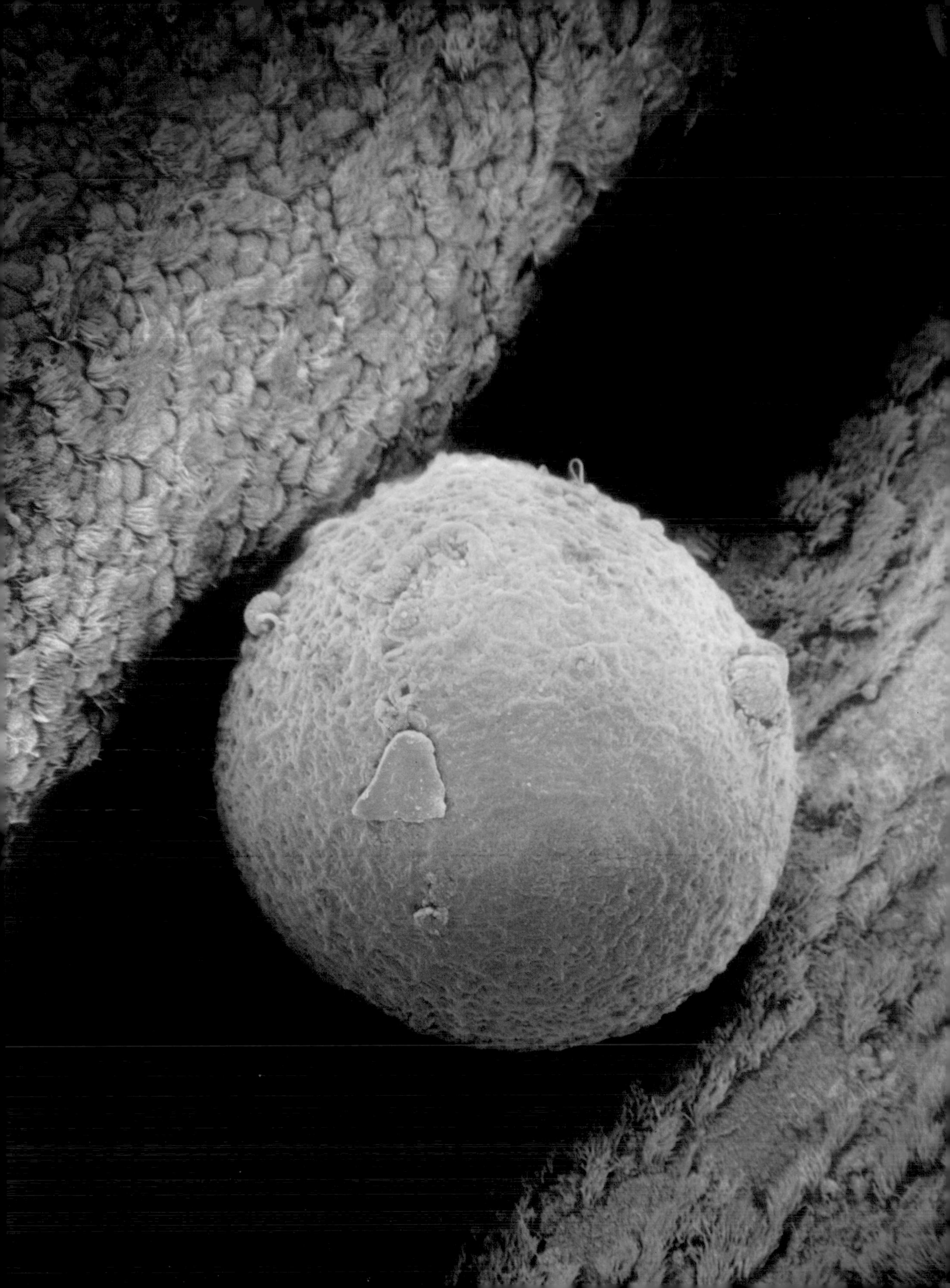

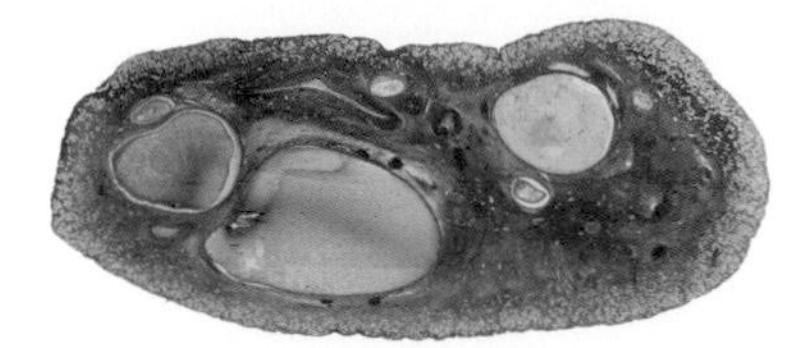

The ovary of a newborn girl with a few nearly mature oocytes.

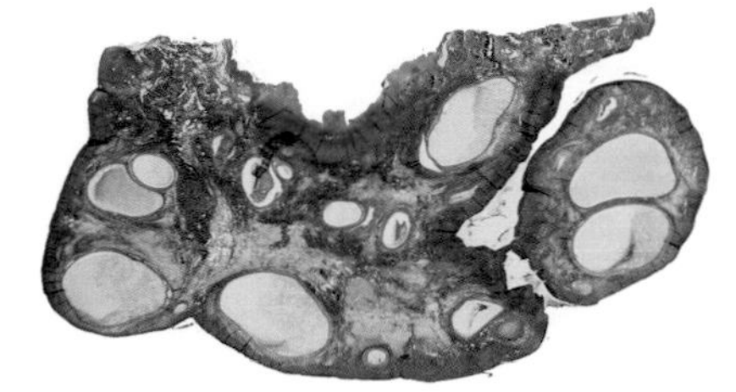

The ovary of a fertile woman.

The ovary of a fifty-year-old woman. No eggs remain.

A century ago young women usually had their first period around the age of fifteen, but today in well-to-do countries such as Sweden and the United States and Britain menarche is much earlier (the average age is 12.5). One explanation is that the timing is governed more by weight than by age. We eat differently today, and girls tend to achieve the critical weight for menstruation (46–47 kilograms/101–103 pounds) far earlier than in the past. Of course, there are wide individual and ethnic variations.

When a young girl's first, often sporadic periods become more regular on a monthly basis, she has probably begun to ovulate, and thus can become pregnant. Early pregnancies are less common than they used to be, but teenage pregnancies are still usual in many parts of the world. In developed countries, the pendulum has swung in the opposite direction. Here, most women have children much later in life than before. Forty years ago, the average Swedish mother would have her first child at age twenty-four. Today the average age of first pregnancy is around thirty, and many first-time mothers are over forty years old. There are, of course, both advantages and disadvantages to having children later in life. While maturity, experience, and job security may help a woman approach the responsibilities of motherhood with more confidence, a somewhat older woman will likely have fewer chances of becoming pregnant or having the number of children she wishes to have. The risk of miscarriage increases as well, since the genetic quality of eggs deteriorates over time.

Inside the egg are mitochondria, little fuel packs that supply the egg with energy and other necessities. There are about a thousand of these mitochondria, which also contain genetic material in the form of special DNA molecules. As a woman ages, the workings of the mitochondria in the eggs deteriorate. This is considered the main reason older women have more difficulty becoming pregnant and miscarry more frequently. If this mitochondrial DNA happens to be damaged, congenital illness may appear in the mother and may be transferred to the baby. Mitochondria are found in all the cells of the body and are also involved in our energy production and aging processes.

> The ovary of a female fetus at week 30, with many small egg follicles and one larger follicle.

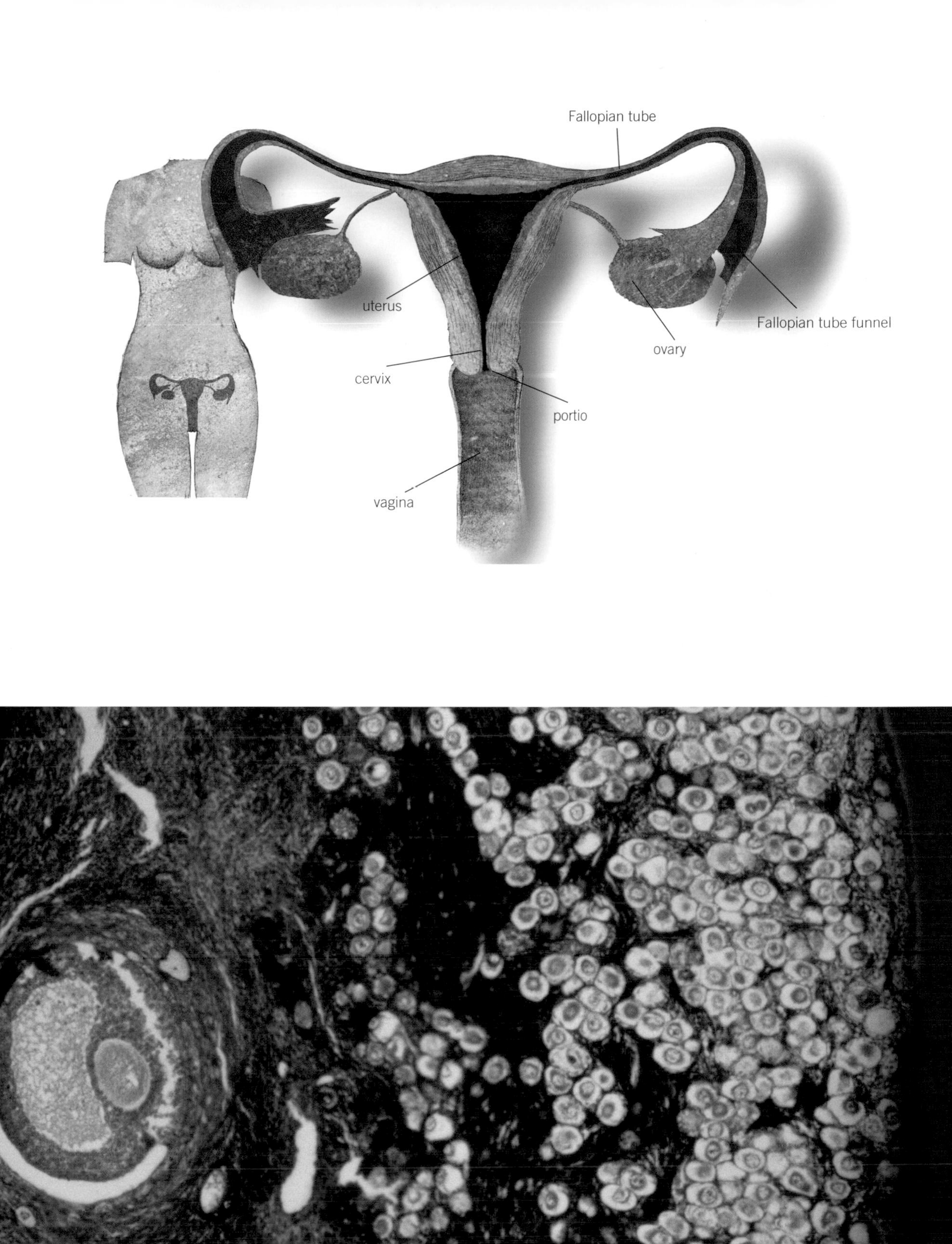
Fallopian tube
uterus
Fallopian tube funnel
ovary
cervix
portio
vagina

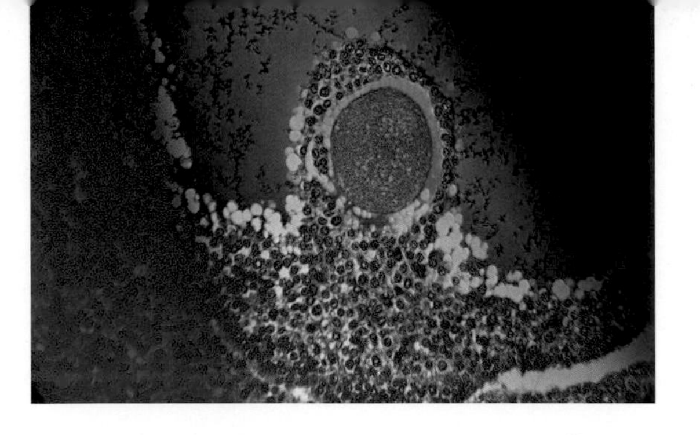

The egg in its follicle just before ovulation.

The menstrual cycle

Every menstrual period in a woman's life begins a new cycle, lasting approximately four weeks. The first hormonal impulses come from the pituitary, a little gland at the base of the brain. Which hormones are to be excreted, and in what quantities, is determined by centers in the lower part of the brain that are in direct contact with the pituitary gland.

The brain has a great influence on the menstrual cycle. Many women, when they are under severe stress or are very worried, experience periods that are delayed or that even stop altogether for one or more months. Anorexia, too, can result in loss of menstruation. Even a small change in a woman's life, such as dieting or vacationing, can result in the temporary disappearance of ovulation and menstruation. So can a very strong desire to become pregnant.

Normal menstruation usually lasts three to five days, during which time the woman loses about 40 milliliters (1.4 fluid ounces) of blood, although most women would probably guess that they lose more. The woman bleeds because the layer of mucous membrane lining the uterus, into which a fertilized egg could have burrowed, is being rejected and replaced with fresh cells for the next cycle. It takes about a week for the membrane to be rebuilt, to achieve the proper thickness, and to develop its fine network of tiny blood vessels.

At the same time, hormones from the pituitary gland have signaled to the ovaries to make a few immature eggs begin to grow and mature. Usually a few of all the possible little clusters react fastest to these signals, and approximately four or five eggs start to mature.

Whether ovulation takes place in one or the other ovary appears to be a random matter. Often the ovaries do not take turns producing eggs with complete regularity. Should a woman need to have one ovary removed, the remaining one will then ovulate every month.

A few days before ovulation a follicle in one of the ovaries begins to develop rapidly. Slowly the follicle rises toward the surface of the ovary.

> Usually just one follicle at a time matures, with just one egg inside. At the time of ovulation the egg rids itself of half its genetic information, creating a tiny cell called a polar body that disintegrates. Twenty-three chromosomes are left in the kernel. The egg is surrounded by a layer of nutrient cells that will be transported along with it after ovulation.

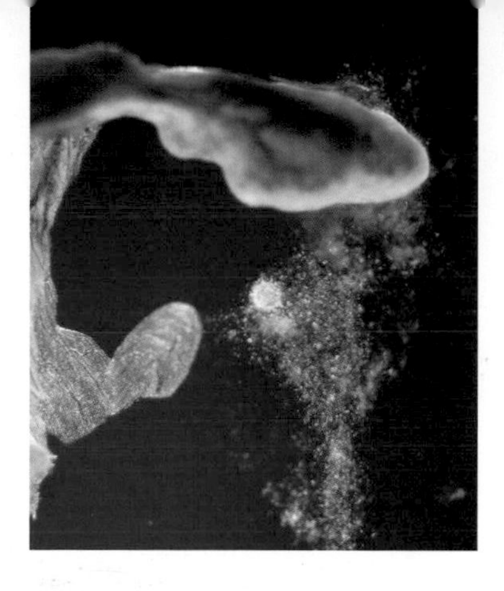

The egg is pushed out of the follicle.

The egg with its nutrient cells.

Ovulation

A woman normally ovulates once a month, about two weeks after her last period. This is the only time when intercourse can result in fertilization. In other words, sperm have the opportunity to accomplish their mission for only a few days every month.

As ovulation approaches, the woman often notices that the amount of mucus discharged from her vagina increases. This mucus comes from the cervix and increases at ovulation, when it becomes crystal clear and stringy. Only when the mucus has this particular quality will the sperm be able to pass up through the cervix. There are other signs of ovulation as well. Some women always feel it in their back, while others may have a day or two of spotting. The woman's body temperature rises by about half a degree just after ovulation. She can check her temperature in the mornings as a way of determining when there is a chance of becoming pregnant.

By about two weeks after the beginning of menstruation, one follicle is approximately 2 centimeters (nearly 1 inch) across or somewhat larger, and it contains an oocyte ready to be fertilized. Suddenly this follicle–which has hitherto enveloped the egg in its protective chrysalis–ruptures, and the liquid that fills it (some 10–15 milliliters/0.6–1.0 tablespoons) is released, along with millions of cells that have been secreting estrogen. In the midst of this swarm is the oocyte itself.

When the oocyte has been released from the follicle, remarkably enough, it is almost always caught up in the supple, ingeniously constructed outer funnel of the Fallopian tube, also known as the oviduct. If a woman has scar tissue from an infection in this part of her body, this sophisticated mechanism is easily disturbed.

A mature egg separates from one of the ovarian follicles and is expelled in a cloud of nutrient cells. Just outside, the funnel of the Fallopian tube is waiting. Should the egg be fertilized while in the oviduct, the nutrient cells can provide it with enough food and oxygen for several days.

< The delicate membranes of the Fallopian tube funnel are ready to catch the egg.

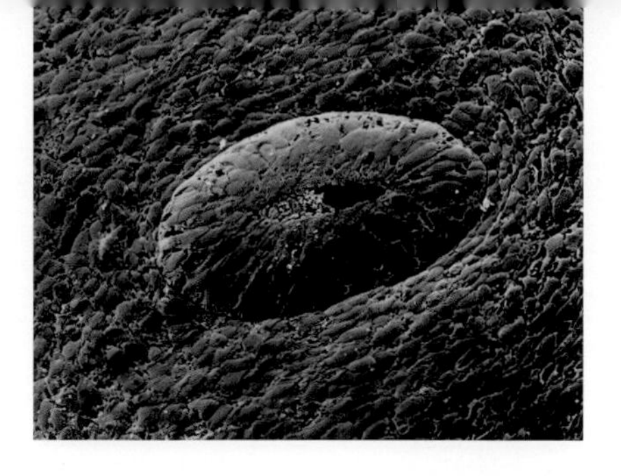

The ovary after ovulation.

Meanwhile the mature follicle, once it has ruptured, goes on to play another important role in the ovary as the corpus luteum. Having secreted large amounts of estrogen, it begins manufacturing a hormone that is chemically related but has very different properties. This is progesterone, which goes out into the bloodstream and influences the outer cell layer lining the uterus, preparing it to receive a fertilized egg.

From the opening of the Fallopian tube funnel, the egg is transported into the sheltered environment of the oviduct itself. There it is prevented from dropping into the abdominal cavity by the beating motions of the minuscule cilia, or hair-like projections, in the mucous membrane lining the tube. The egg remains in this open, wider part of the oviduct in anticipation of possible fertilization for about forty-eight hours. It does not always lie still but rolls slowly along the membranes, awaiting its male partner. In the snug environment of the oviduct, the egg and its cloud of nutrient cells come to full maturity.

Fertilization takes place in the Fallopian tube. It makes no difference whether the sperm or the egg gets there first. When the woman ovulates, sperm may be lying in wait in the folds of the mucous membrane lining the tube, where they can survive for several days. An egg is more sensitive: once it is in the tube, the sperm have to appear—and fertilization must take place—within twenty-four hours if there is to be a viable chance of pregnancy.

If no sperm arrives, or if the egg's encounter with the sperm fails to result in a viable embryo, it will simply continue down through the oviduct into the uterus and then out past the cervix, to be discharged through the vagina. Then in about ten days, the woman will menstruate.

The follicle has collapsed; a hole gapes in its middle from which the oocyte was released. It is now transformed into a corpus luteum, and its cells begin producing progesterone.

> The egg rolls around in the Fallopian tube, bouncing softly off the thousands of tiny cilia that move it gently along its course. The surrounding layer of nutrient cells begins to dissolve, owing both to friction and to the effect of enzymes in the tube's secretions (overleaf).

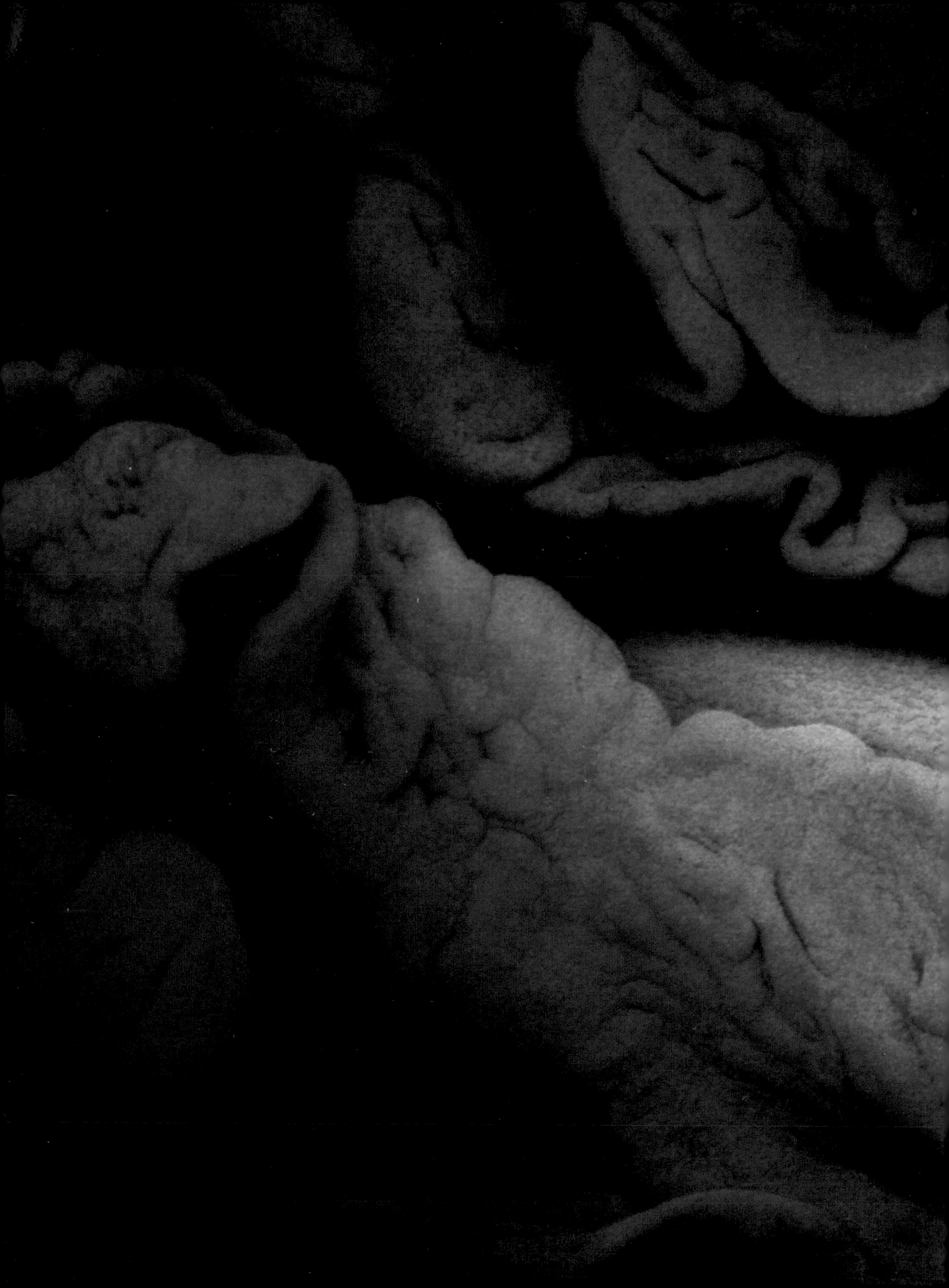

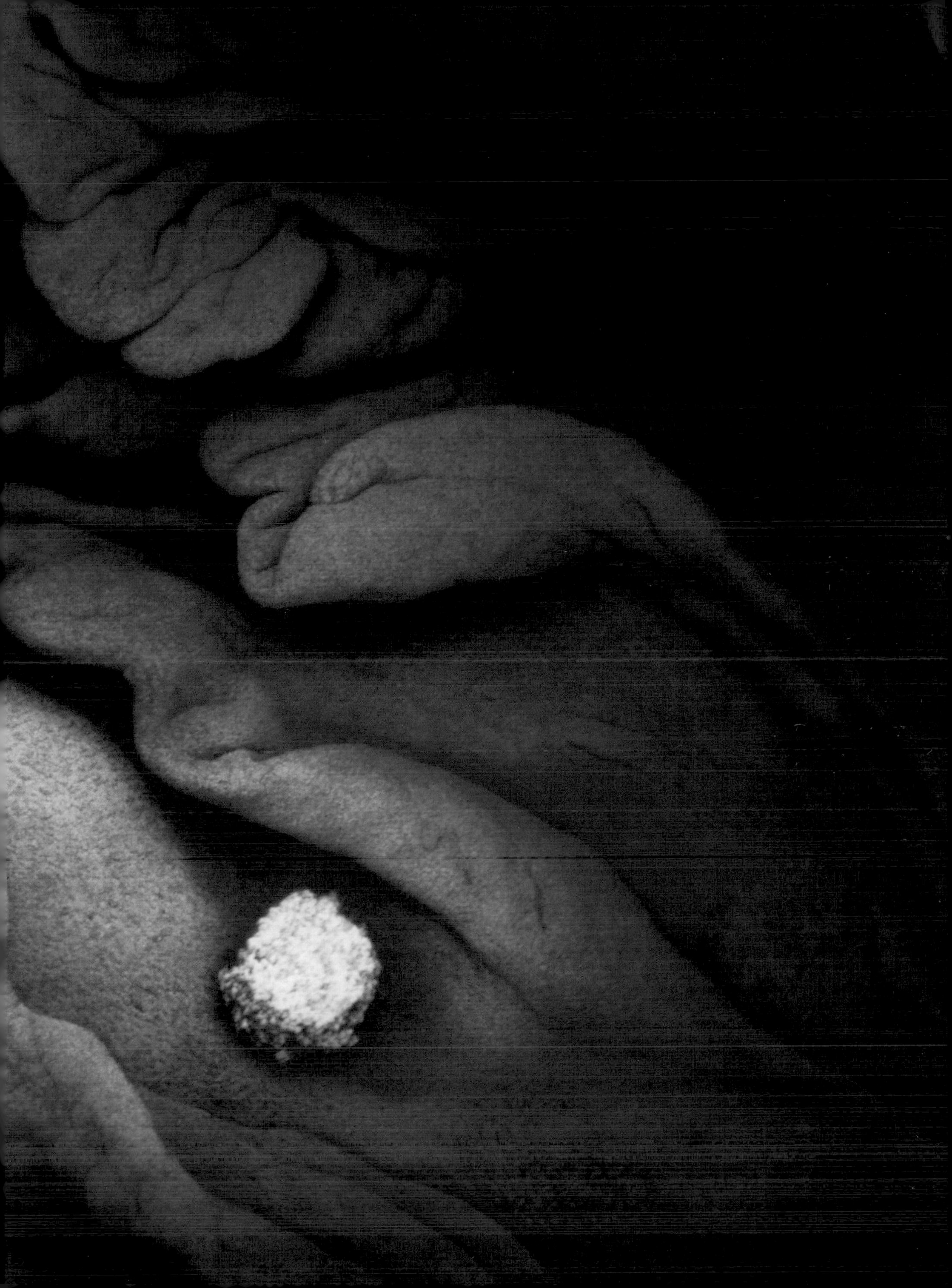

The expectant father

Sperm, or male sex cells, are much smaller than eggs. A sperm consists of a head, which contains the genetic mass, a midsection, and a long, thin tail. The mission of the sperm is to deliver the man's genetic material to the egg.

Primitive sperm cells, known as spermatogonia, are already in place in the testicles of a newborn baby boy. Hormones from the pituitary govern the sexual development of boys and trigger production of viable sperm. Why do boys become sexually mature—that is, begin to produce sperm capable of fertilizing an egg—at the age of about twelve or thirteen? There is not yet a clear answer to that question, but we do know that before that time the thymus gland impedes the sexual maturation process, which is then initiated by a complex interaction of many factors, including adequate nutrition supply and various growth hormones and hormones from the adrenal cortex. Genetic factors also play an important role.

Hormones, particularly testosterone, the male sex hormone, govern boys' bodily growth, muscle structure, development of external sexual organs, voice change, whiskers, and development of body hair. Luteinizing hormone (LH) helps the testicles to produce testosterone, and follicle stimulating hormone (FSH) triggers both sperm production and the maturation of the sperm, making them viable for fertilization. Both LH and FSH are produced by the pituitary. The interaction among the hormones is delicately balanced; disturbances may result in permanent damage.

Sperm in the seminiferous tubule (seen here in cross-section). The genetic information is tightly packed in the head of the sperm.

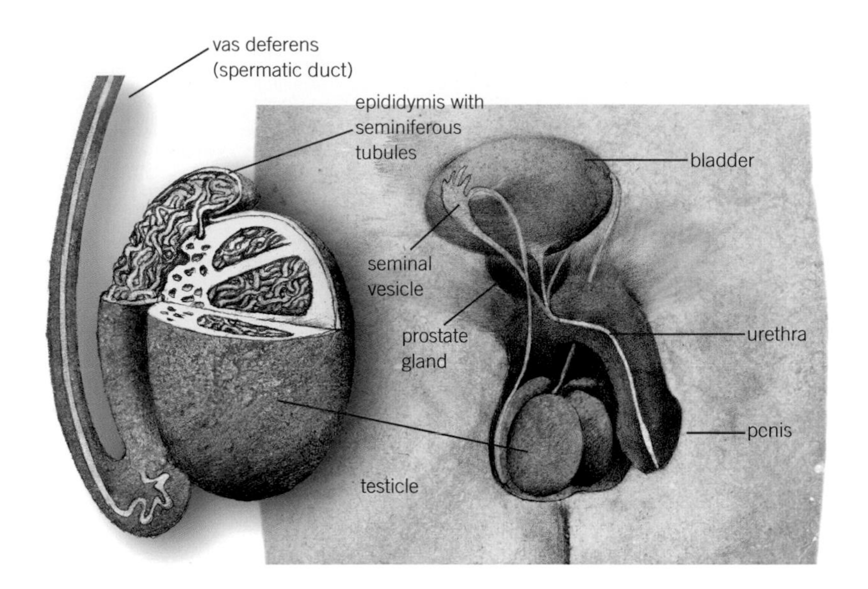

Inside the sperm factory

While a woman generally develops only one egg per month, a man can produce millions of new sperm every day. Every time he ejaculates, his body produces 2–6 milliliters (0.3–1.0 teaspoons) of seminal fluid that can contain more than five hundred million sperm, sometimes up to one billion. This production continues in some men until the age of eighty, although the speed and quality of sperm production gradually decline.

The sperm production process appears to be considerably less efficient than women's egg production. Many sperm lack either adequate propulsion ability, endurance of movement, or—worst of all—correct genetic information. Over 90 percent of the sperm a man produces may be defective in some way, without negative effects on his fertility; the sperm capable of fertilizing an egg are still plenty in number.

The body takes over seventy days to make a sperm cell. Sperm are produced in the testicles, in a meandering tangle of canals called seminiferous tubules several hundred meters long. In the walls of these canals, the sperm begin to grow.

Sperm production begins at the outer edges of the seminiferous tubules, where immature sperm are stored. An immature sperm contains forty-six chromosomes. Early in the production process it divides itself into two spermatids, which have twenty-three chromosomes each so as to be able to unite with the twenty-three chromosomes of an egg, creating a new life.

Seminiferous tubule (shown here in cross-section). Each tubule has a central channel in which the newly produced sperm lie waiting to be transported to the epididymis.

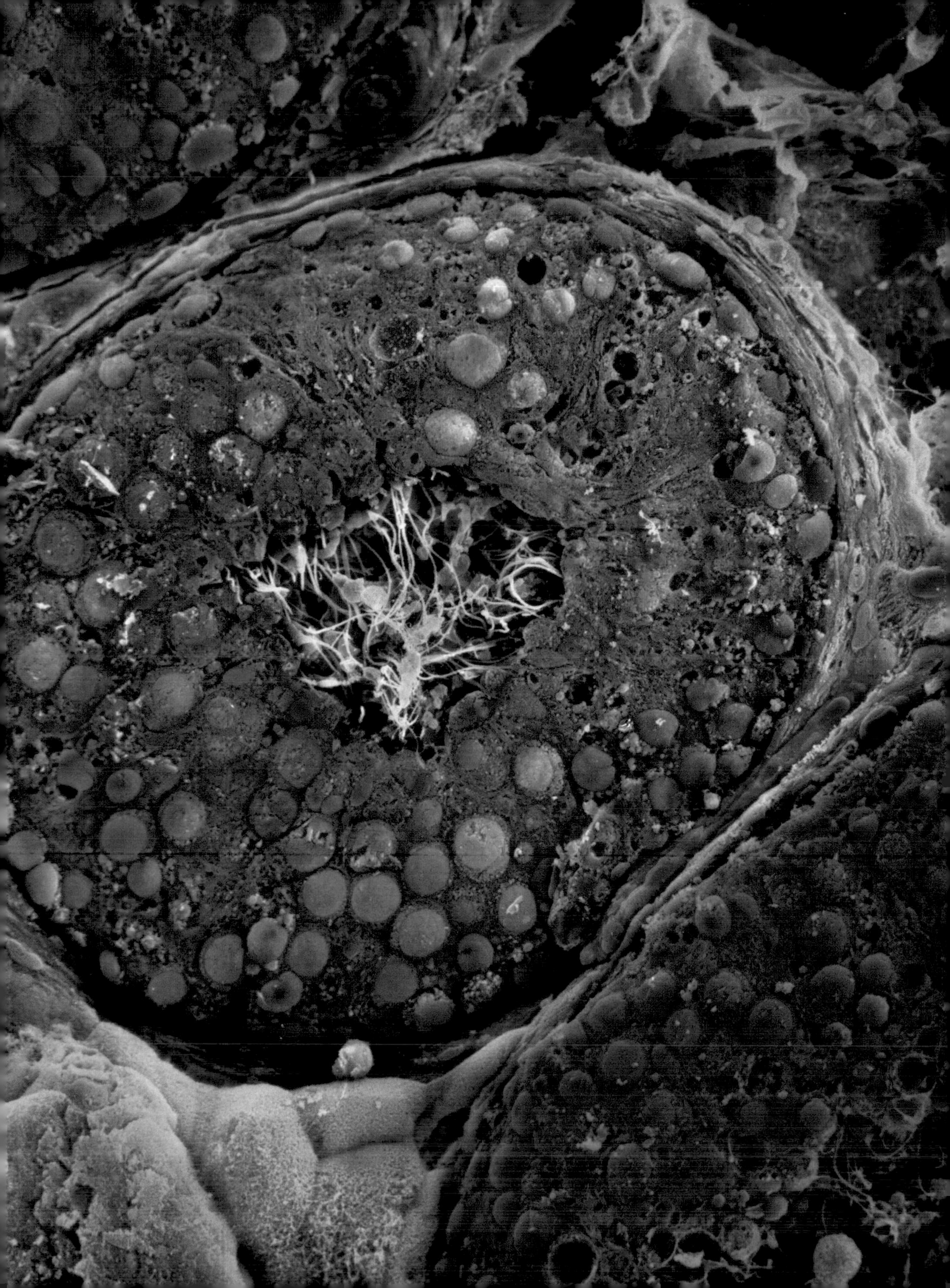

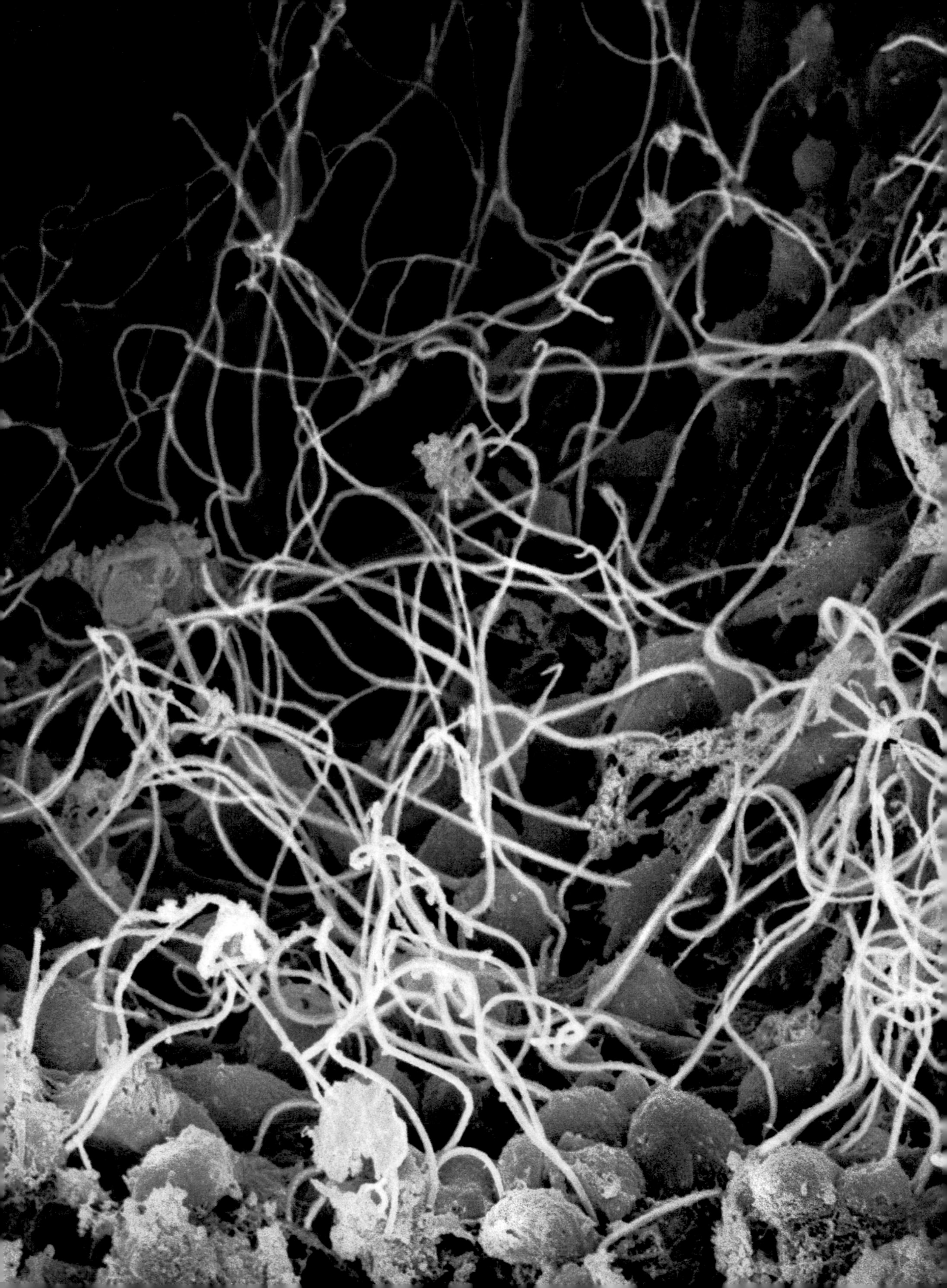

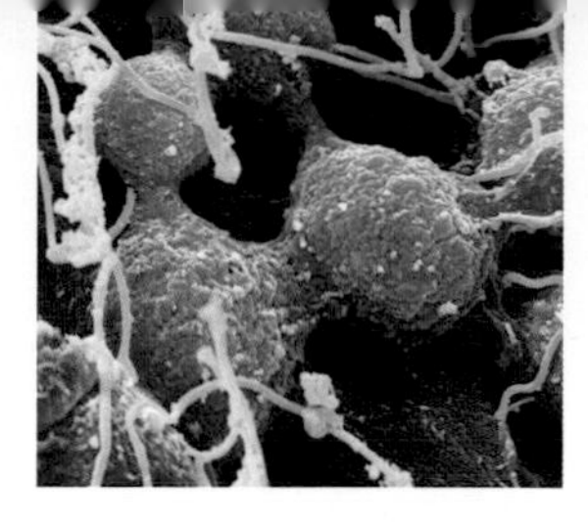

Immature sperm dividing.

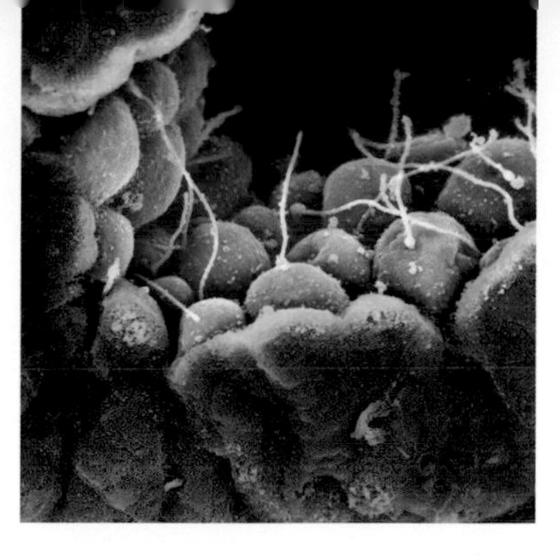

Partially developed sperm.

The spermatids quickly begin to develop tails and they move from the walls into the central channel of the seminiferous tubule, tail first, head last. As a sperm approaches maturity, its appearance changes. The rounded cell becomes more streamlined, a shape better suited to the task at hand: swimming competitively through a woman's reproductive system. For now, the sperm are still unable to propel themselves, and are pulled passively along in the secretion that flows through the tubules.

These tubules meet and join, merging into wider channels, and eventually the sperm reach the central sperm holding tank known as the epididymis, where the final maturation process takes place. There the tail begins to function and the sperm become mobile. Sperm do not survive for long. If they are not ejaculated, they eventually die, making way for newly produced sperm.

Scientists in several countries have noted an alarming decrease in human males' sperm count. Intensive research into the reasons is ongoing; they may include air pollution and environmental toxins in the food chain. Stress and toxins in certain work situations are also considered possible causes. The treatment of livestock with antibiotics and hormones that then remain in the meat we eat may also be a factor. Anabolic steroids and other testosterone-like chemicals can temporarily reduce fertility, but are not believed to have a lasting effect.

When immature sperm divide, two spermatids are formed with twenty-three chromosomes each. Seen at right are large numbers of half-developed sperm in the process of maturation. Most already have tails.

< Still in the central channel of the seminiferous tubule are hundreds of sperm with tails that have grown ever longer; they seem to risk becoming tangled up. A slow steady flow of fluid transports the sperm to the catchment area in the epididymis, where the tails finally become capable of propulsion.

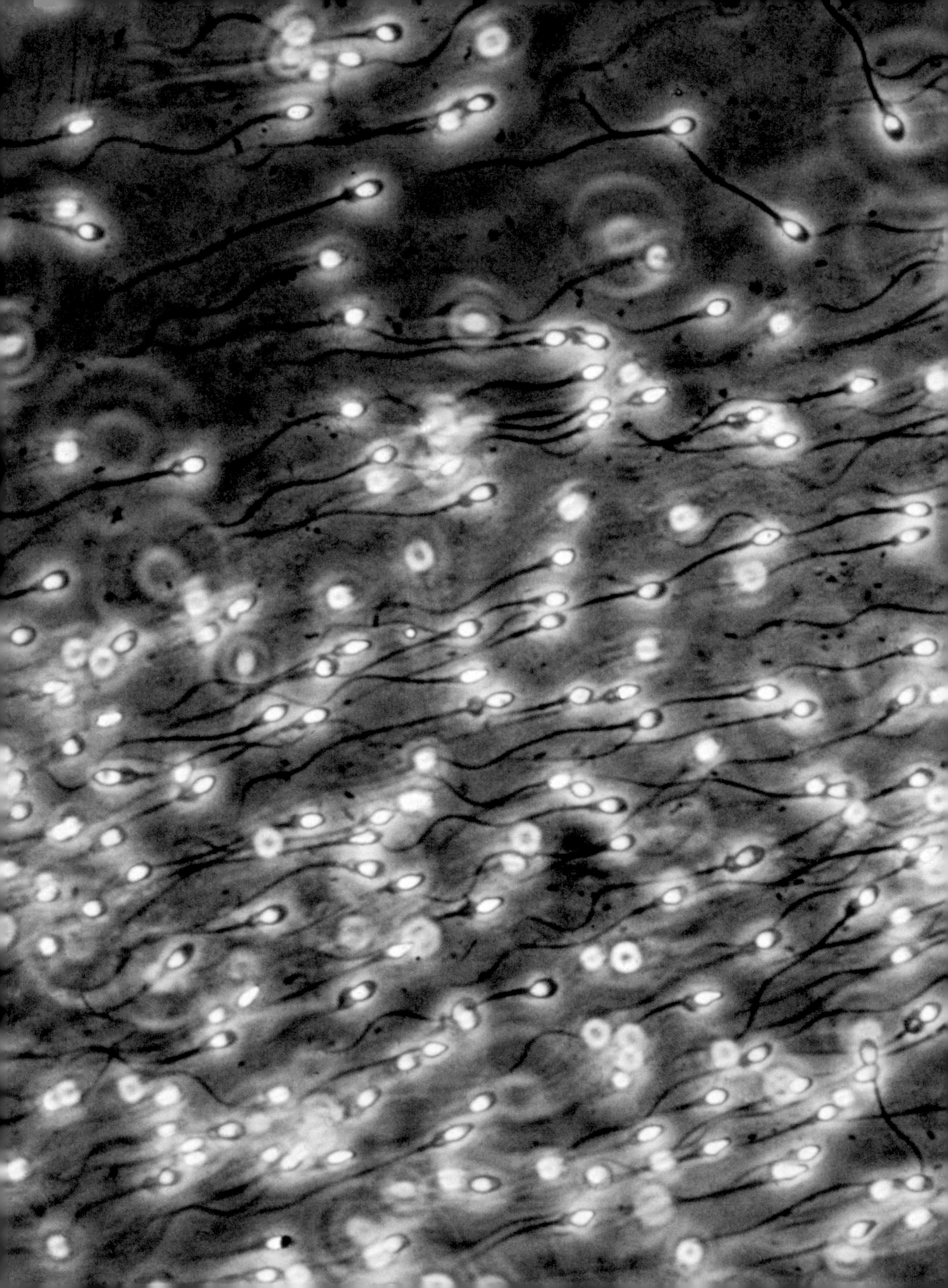

FERTILIZATION AND CONCEPTION

For each of us, life's journey began with the meeting of egg and sperm. Nearly all the information needed to make us into the unique individuals that we are was already there in that moment, stored and waiting. But procreation is not a simple process. A complex jigsaw puzzle of events must fall into place, each piece fitting perfectly, to create a new life able to take root and grow in the shelter of the womb.

Journey toward the egg

During intercourse, when the man's pelvic muscles contract in orgasm, his sperm shoot out of the epididymis into the urethra, mixing with secretion from the prostate gland. This secretion contains substances that enable the sperm to move more effectively toward the woman's egg. His seminal fluid shoots out into the vagina: the sperm and secretion from the prostate first, followed by a gelatinous fluid that flows from the seminal vesicles to the urethra. In an erection, the blood flow through the penis increases, and the erectile tissue (the corpora cavernosa) fills with blood and expands. After ejaculation, the blood flow subsides, and the tissue returns to normal size.

Some of the sperm that enter the vagina flow right back out again, while others rest on the mucus, and in the many folds of the mucous membranes. Their tails propel them as they drive deeper into channels in the mucus. These passages are often narrow and congested. During ovulation and for the following few days, the woman's estrogen helps unblock these channels, enabling the sperm to proceed.

The cervix, protruding down into the top of the vagina, is the eye of the needle that the sperm have to get through on their way toward the egg. At the time of ovulation the vagina produces a large amount of a crystal-clear discharge, the function of which is to screen the sperm and perform the first selection process. Sperm that are too weak become lodged in this mucus. Others make their way through and swim on through the narrow mucous channels of the lower cervix. The portio, or entrance to the uterus, is open only a few days out of every month, after which it is plugged shut again by thick mucus that shuts out not only sperm but also all kinds of bacteria.

The sperm have a long, laborious journey toward the egg, and as they travel through the cervix and the uterus and up into the uterine tube, most succumb. Only a fraction of them reach their goal. Sperm swim into both Fallopian tubes as they career blindly ahead, although the one Fallopian tube that contains the egg is often a little more open and welcoming. Thus some sperm "guess wrong" and end up in the tube that has no egg that month.

(pp. 38–39) After ejaculation a swarm of sperm enters the vagina and takes off for the cervix. This is the initial heat in the race for life, with up to five hundred million competitors.

> After their long swim, the sperm have finally arrived at their goal. The egg is still surrounded by its cloud of nutrient cells; a number of the sperm immediately begin to work their way past these cells and through the outer shell of the egg.

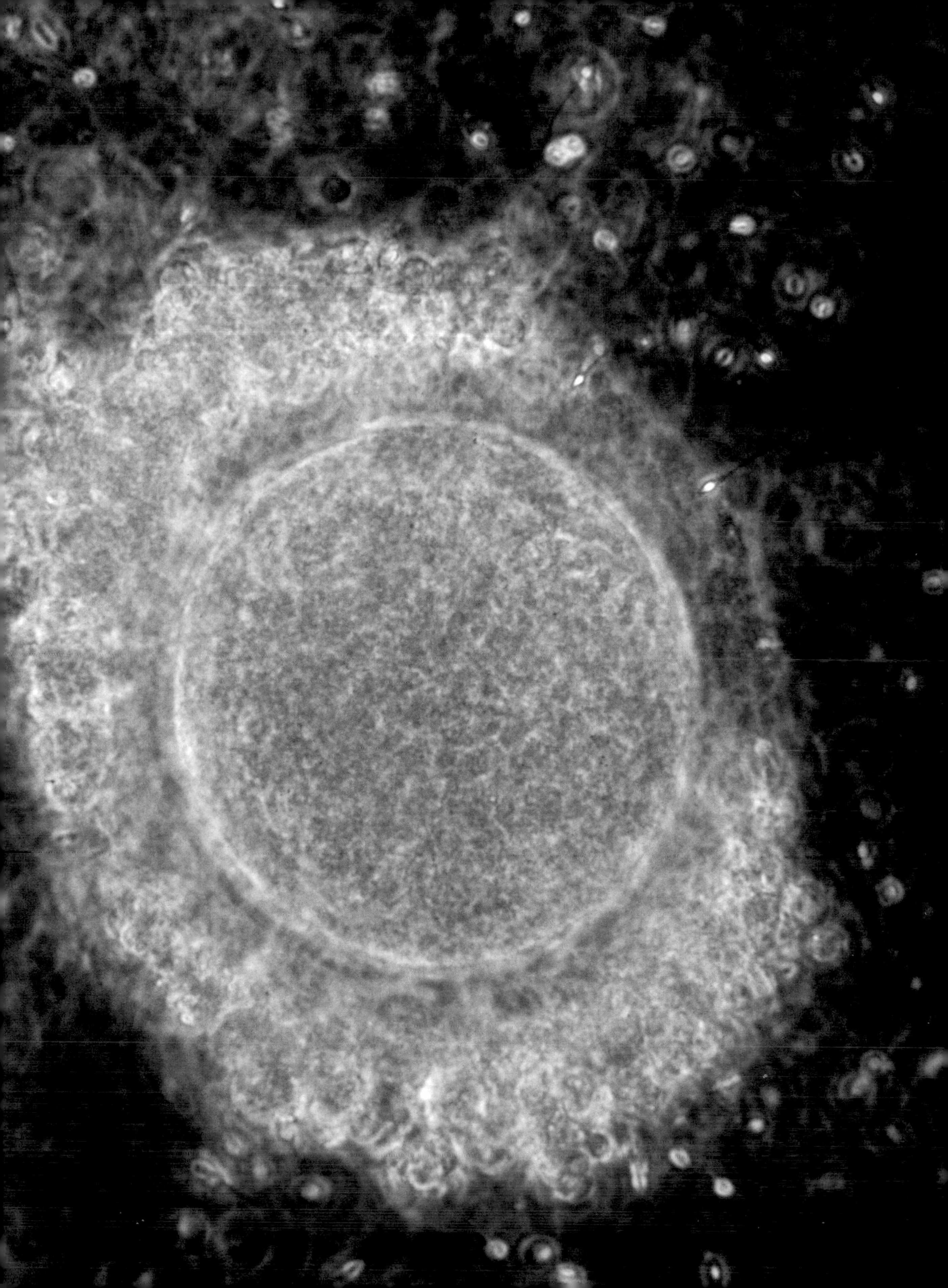

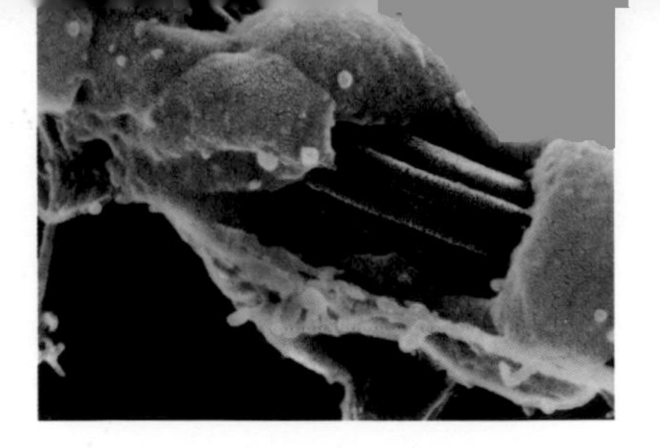

Microtubules in a sperm tail.

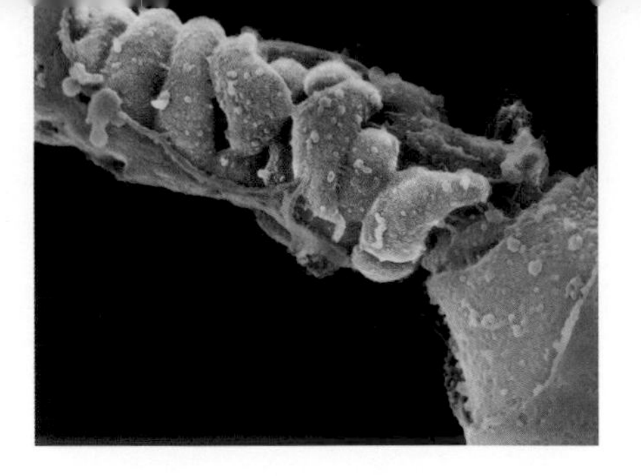

Mitochondria just behind the head of the sperm.

Making headway through the hair-like cilia that line the cervix and Fallopian tube is not easy. Many sperm get stuck in the corrugated landscape, or find they lack the strength to continue. The cilia, moreover, beat "against the flow," making the journey for the sperm even more of an uphill struggle. Some stop and rest along the way, and it is thought that some sperm, particularly those from young, healthy men, can remain alive and potent for up to four to five days after intercourse.

To accomplish this demanding journey, the sperm require energy. Just behind the head of each sperm is a midsection containing a packet of energy in the form of mitochondria. It has been estimated that a thousand swishes of the tail propel a sperm approximately one centimeter and that the energy housed in the midsection of a sperm is sufficient to allow it to swim for hours. The mitochondria can also refuel by extracting sugary substances from the environment. The wriggly tail that propels the sperm forward is a system of microtubules, a brilliant construction not unlike an electric cable in cross-section. Without these tubules the sperm would not be able to swim.

During the journey the sperm change successively, under the constant influence of substances from the woman's reproductive system. Glands among the clumps of cilia produce a secretion that helps the sperm mature. At one point they are "capacitated," after which they are capable of fertilizing the egg.

It usually takes several hours for a sperm to make its way through the vagina and into the uterine tube, a journey of 15–18 centimeters (6–7 inches). But under favorable conditions some quick swimmers are able to reach the uterine tube within half an hour. Researchers speculate that sperm transport may be facilitated by the female orgasm. Still, it is not certain that any one of these fastest-swimming sperm will be the one to fertilize the egg.

Cable-like microtubules in the sperm tail provide propulsion, and mitochondria just behind the head give the sperm the energy it needs to make the journey to the egg. Fertilization usually takes place within 24 hours.

> Lots of sperm use up all their energy in a futile struggle with the dense mass of cilia.

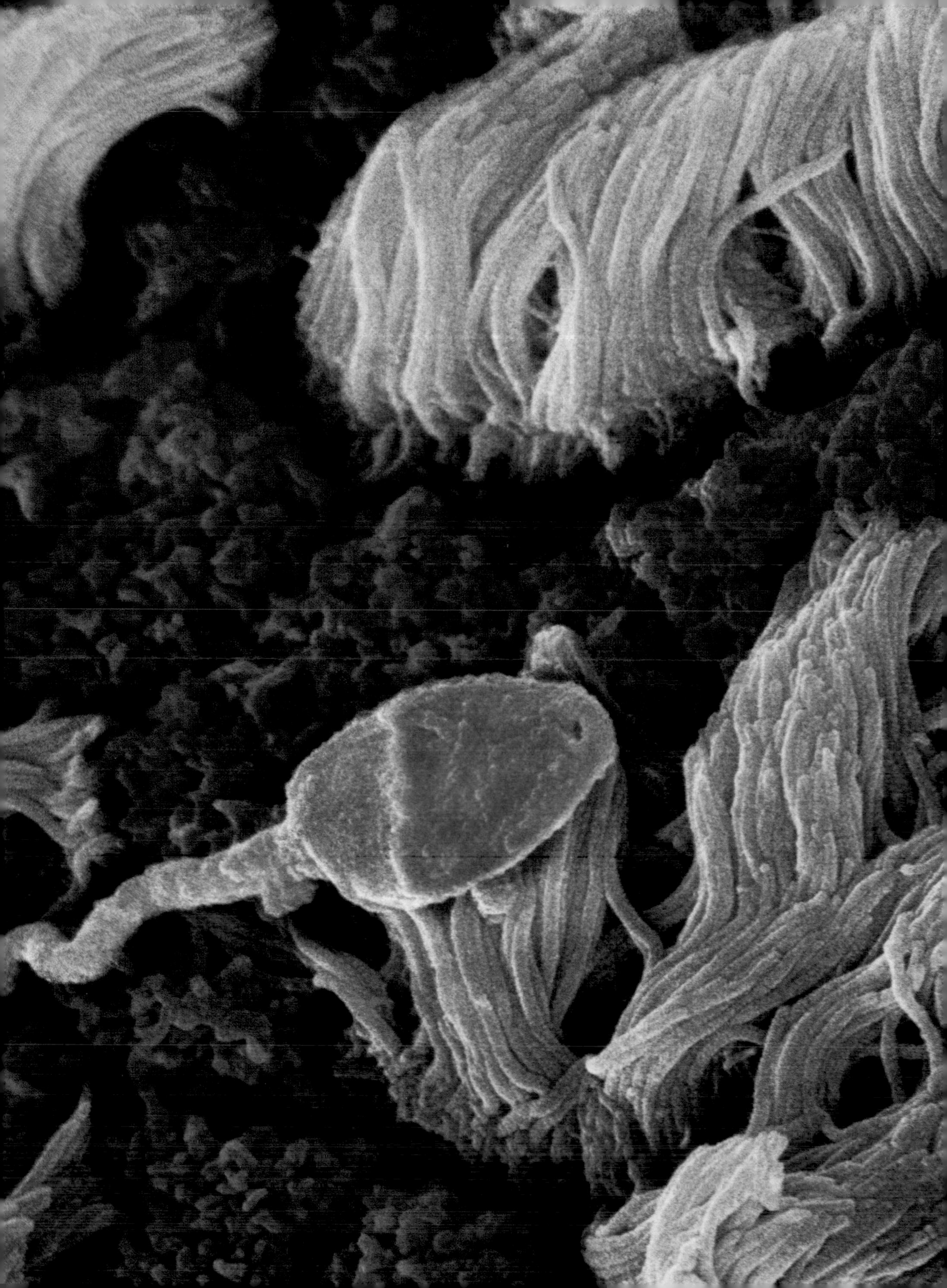

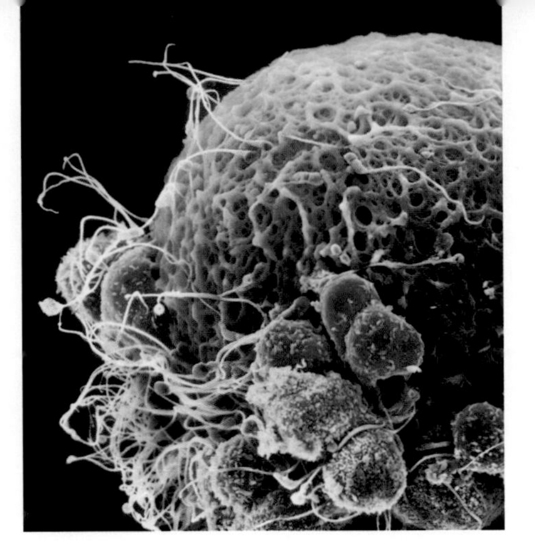

The sperm clear away the nutrient cells around the egg.

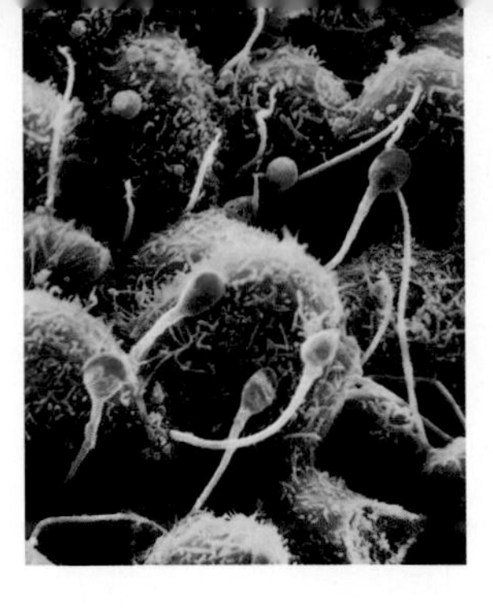

Sperm with acrosomes (in red).

Exposing the egg

When the first sperm have reached the egg, the actual fertilization process begins. The sperm swarm around the large sphere, beginning their frantic efforts to force their way inside.

The egg is still enclosed in nutrient cells, although a number of these, having fulfilled their task, have detached from the egg during its passage through the Fallopian tube. At any given instant, there may be a hundred or more sperm fighting to reach the egg. As each sperm pushes through the cloud of nutrient cells, a chemical reaction dissolves its acrosome, an organelle that has covered and protected its head, and enzymes are released that help the sperm to dissolve the cells surrounding the egg. Many sperm die during this process, but they pave the way for the eventual winner. After a few hours, the surface of the egg is visible.

The interior of the egg is shielded by a strong shell that is difficult to penetrate and consists not of cells but of a tough, almost hard cohesive material. The sperm will have to penetrate this shell to enter the cytoplasm of the egg, the nucleus where the woman's genetic information is stored.

Each sperm functions rather like a drill: the movement of the tail propels the head around and around like the bit of the drill. The eggshell is so thick and so hard that it is surprising that the sperm are able to penetrate it, but it doesn't take long for the sperm to find little nooks and crannies that facilitate its way in. Before long, a number force their way through.

Before they can enter the egg, the sperm have to remove the remaining nutrient cells from around its shell. The caplike acrosome protecting their heads dissolves, releasing enzymes that aid in the disrobing. Here, the surface of the egg is already partly visible.

Some 20,000 lashes of its tail have finally propelled this sperm to the egg. It has begun to penetrate the shell but has not quite reached the interior.

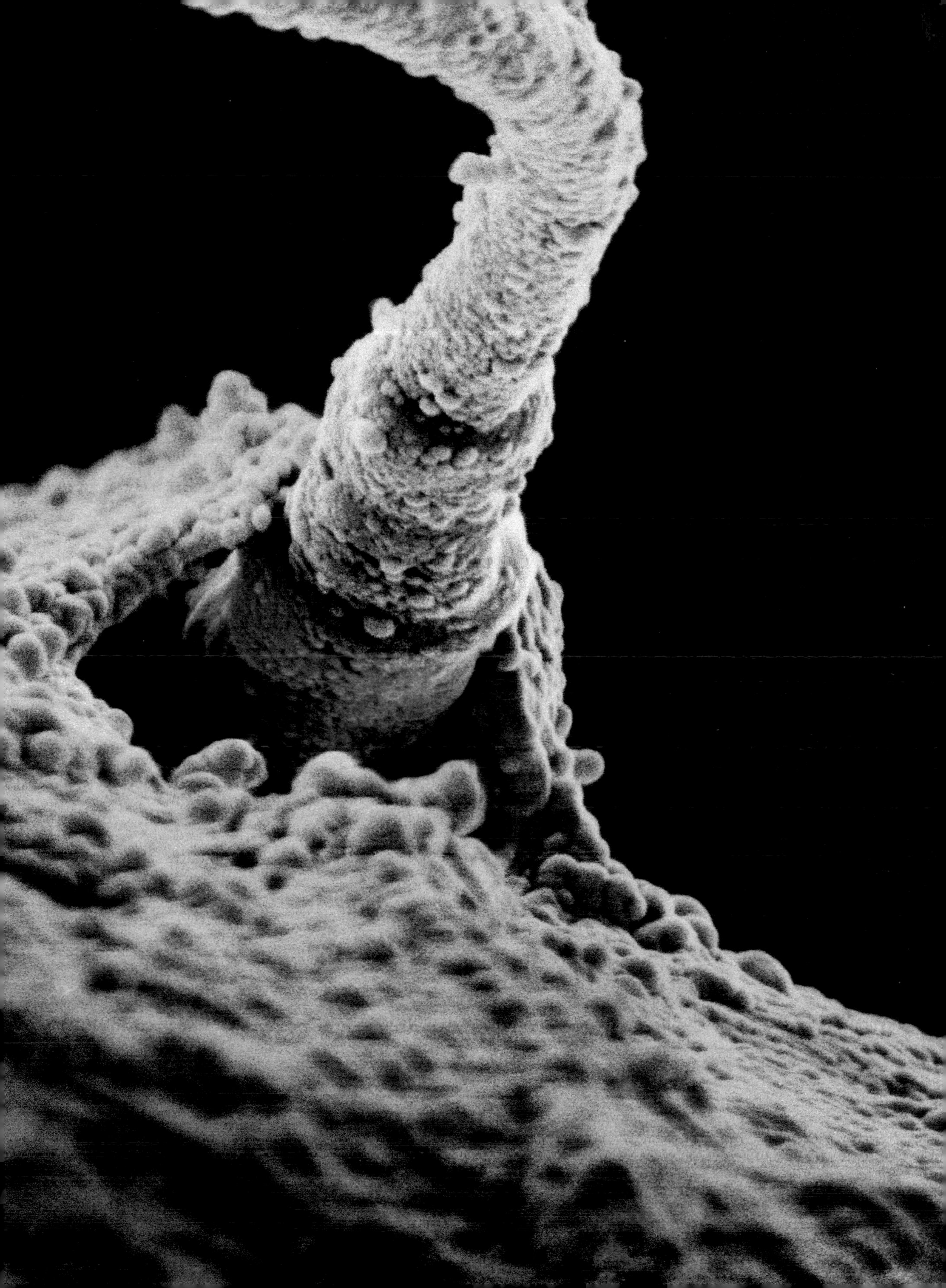

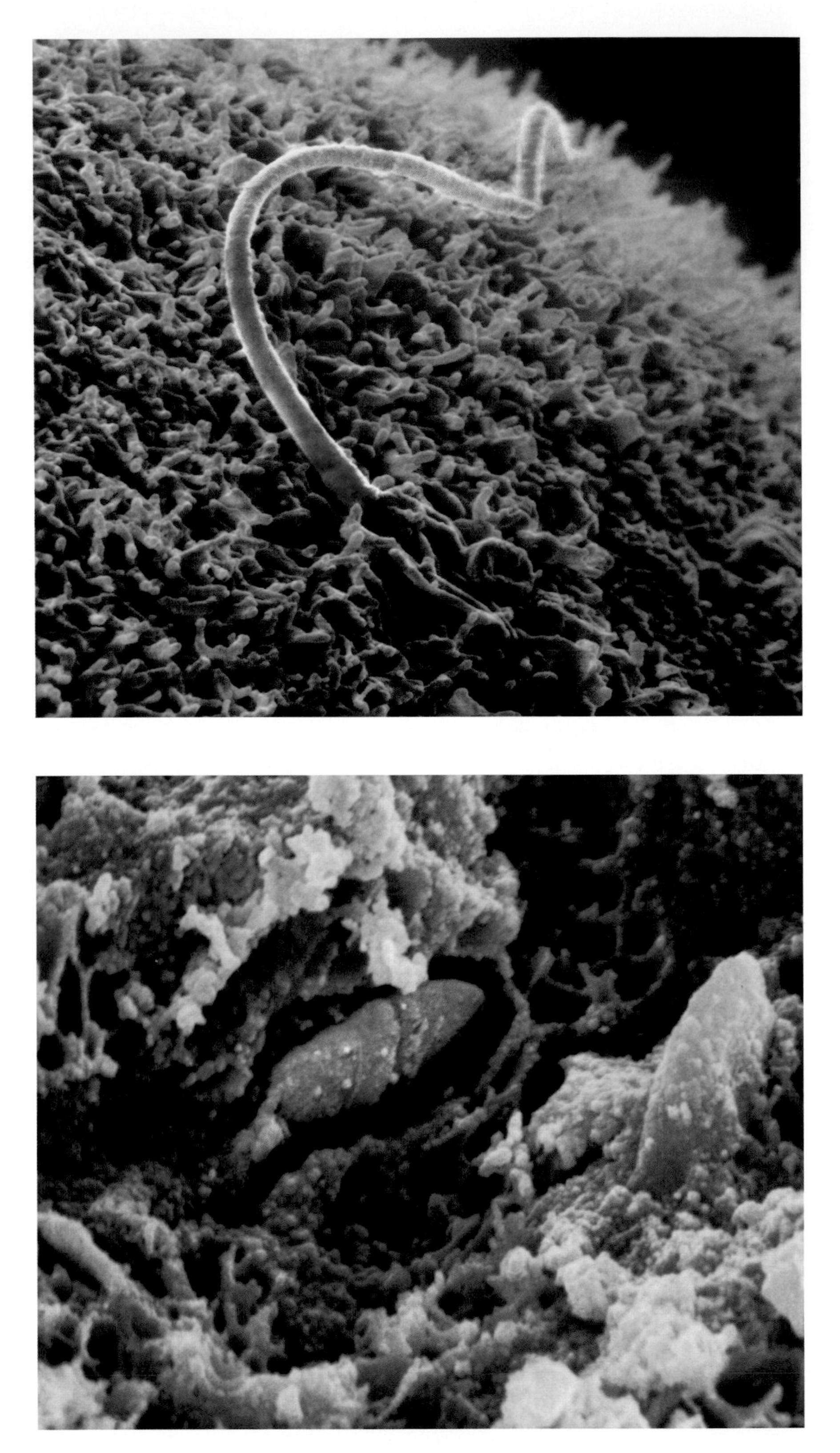

Several sperm reach the narrow gap between the outer shell and the protective inner membrane of the oocyte—but only one of them can fertilize the egg.

< The winning sperm has penetrated the cytoplasm of the egg, where the woman's genetic information is stored.

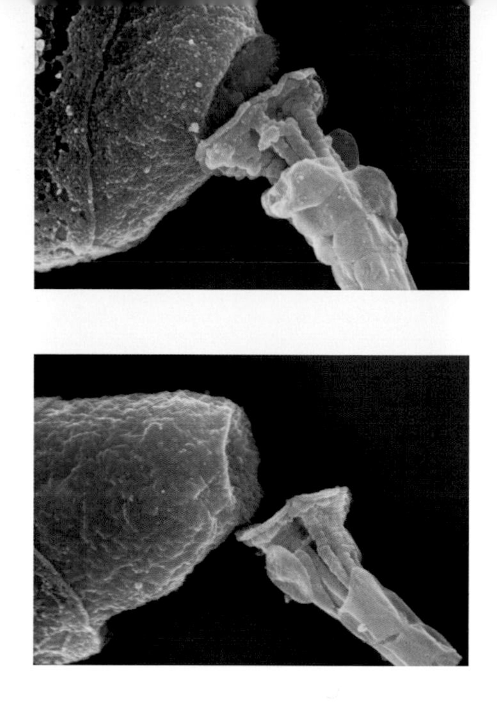

The tail of the sperm breaks off.

After breaching the shell, the sperm enter a fluid-filled space between the shell and the egg itself. The passage through the shell removed the last vestiges of the acrosome. Now, finally, they have the chance to penetrate the egg.

As the sperm congregate in this space, one sperm suddenly stops and latches onto the membrane surrounding the egg cell. Its head and midsection disappear into the interior of the egg, followed within a few moments by the rest of its body. Once inside, the sperm head swells up, becoming a distinct nucleus carrying the man's genetic material. As a spaceship loses its rocket booster, the sperm loses its tail, as well as the expired "engines" in the midsection that have now served their purpose.

As soon as the sperm penetrates the egg's cell membrane, something amazing happens. The chemical composition of the egg undergoes a sudden change, as a rapid stream of ions modify the electric current across its membrane. The other sperm can no longer enter. This is essential, because if the egg were to be fertilized by more than one sperm, the genetic information would be disturbed, and development would stop.

Once the head of the sperm has entered the egg, the tail is no longer needed. It has done its job.

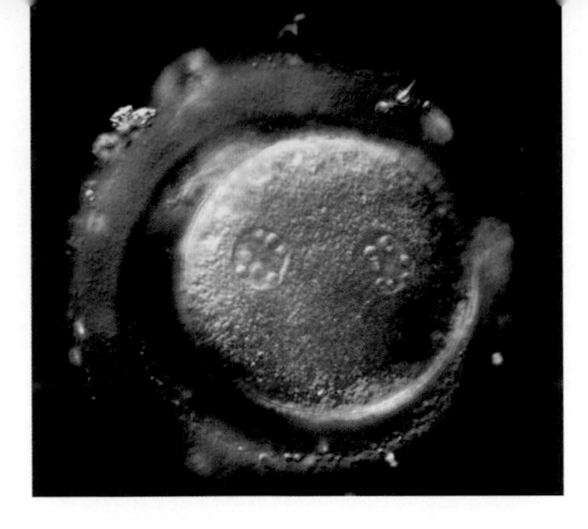

A fertilized egg with two nuclei.

A new human being is created

At about the same instant that the sperm head swells up into a nucleus inside the cytoplasm, another nucleus is forming inside the egg, in the spot where all the genetic information from the woman has been stored. Now these two nuclei must meet. To help them, spiderweb-thin filaments of tubulin develop from a part of the sperm head called the centriole and stretch across the interior of the egg. Like train tracks, the filaments guide the nuclei through the egg cytoplasm, ensuring that all the right pieces of genetic material meet and fuse.

Powerful kneading movements of the egg cytoplasm help bring the two nuclei closer. Soon, the nuclei meet and fuse. At that moment a unique genetic code, a human embryo, is created. Some of its genes have come from the mother, some from the father. The potential variety of combinations is virtually infinite.

After the nuclei fuse, their outer walls dissolve and they are incorporated into the cytoplasm of the egg. Inside the eggshell there is now just a single cell, the original cell for all the billions of cells that will develop into the body of the future human being. Fertilization is complete.

Two cell nuclei of approximately the same size are now in the cytoplasm of the egg. One is the distended head of the sperm, containing the man's genetic information; the other is the nucleus of the oocyte, containing the woman's genetic information. They move slowly toward each other and toward the center of the egg.

> Deep inside the egg, the miracle occurs: the male and female nuclei fuse, forming new chromosome pairs and, with them, an entirely new individual.

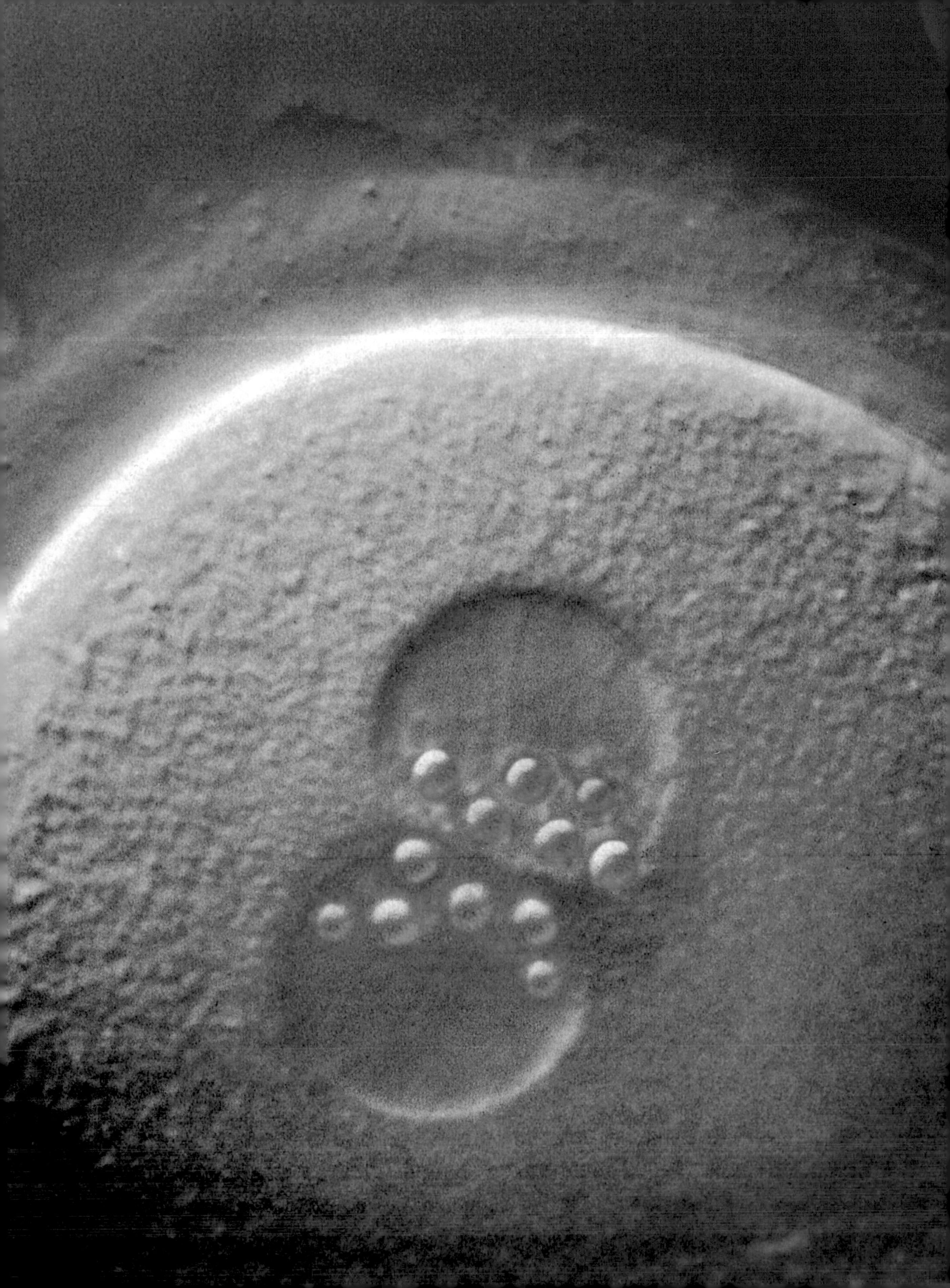

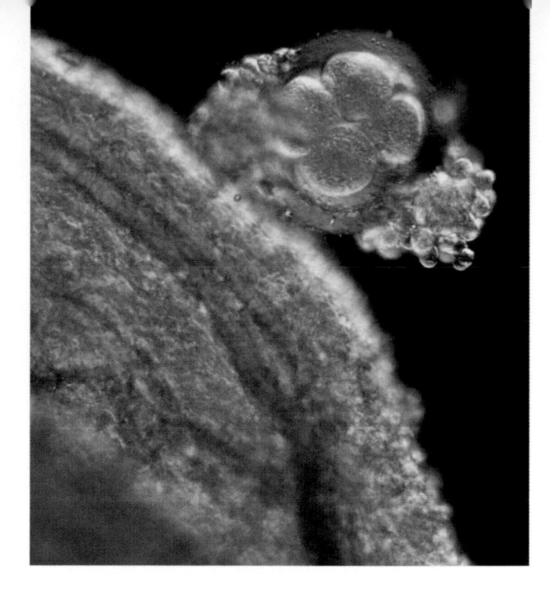

A fertilized egg in the Fallopian tube.

Journey through the Fallopian tube

A few hours after the fusion of the two nuclei, the egg divides for the first time, with great energy, creating a precise copy of itself. Such cell division is the cornerstone of all life in the universe, the key to the miracle of creation. The centriole helps to distribute the genetic material during this first division. The egg now consists of two cells, each bearing genes from both mother and father.

The environment of the uterine tube is perfectly adapted to the requirements of the newly fertilized egg (the zygote). Nutrients flow through the mucous membrane, and the tiny amounts of waste products produced are diluted in the sea of fluids surrounding the egg.

While it is being fertilized, the egg rests in the folds of the mucous membrane of the Fallopian tube, rocking slowly back and forth, following the movements of the woman's body. A few days later it begins its journey down to the uterus, swept forward by the millions of tiny cilia that cover the mucous membrane, all beating in the same direction. The wall of the tube consists of muscles that can contract, conducting the flow of fluids in the same direction. In other words, the Fallopian tube is constructed to prevent an egg from accidentally slipping into the abdominal cavity.

The egg travels through the uterine tube for another two or three days, dividing several times, initially at twelve- to fifteen-hour intervals. Forty-eight hours after fertilization, there are four cells, and twenty-four hours later, eight.

< The egg has divided for the first time, forming two identical daughter cells.

Cell division continues: four visible cells, perhaps eight in total, make their way down the cilia-lined Fallopian tube toward the uterus. The little clump is still surrounded by some of the nutrient cells that nourished the oocyte.

Outside the egg, sperm are still struggling valiantly, whipping their tails in vain, in hopes of fertilizing the egg. But now that there is a winner, the egg has effectively sealed itself off, preventing additional sperm from entering. These sperm may take up to several days to give up the struggle and die.

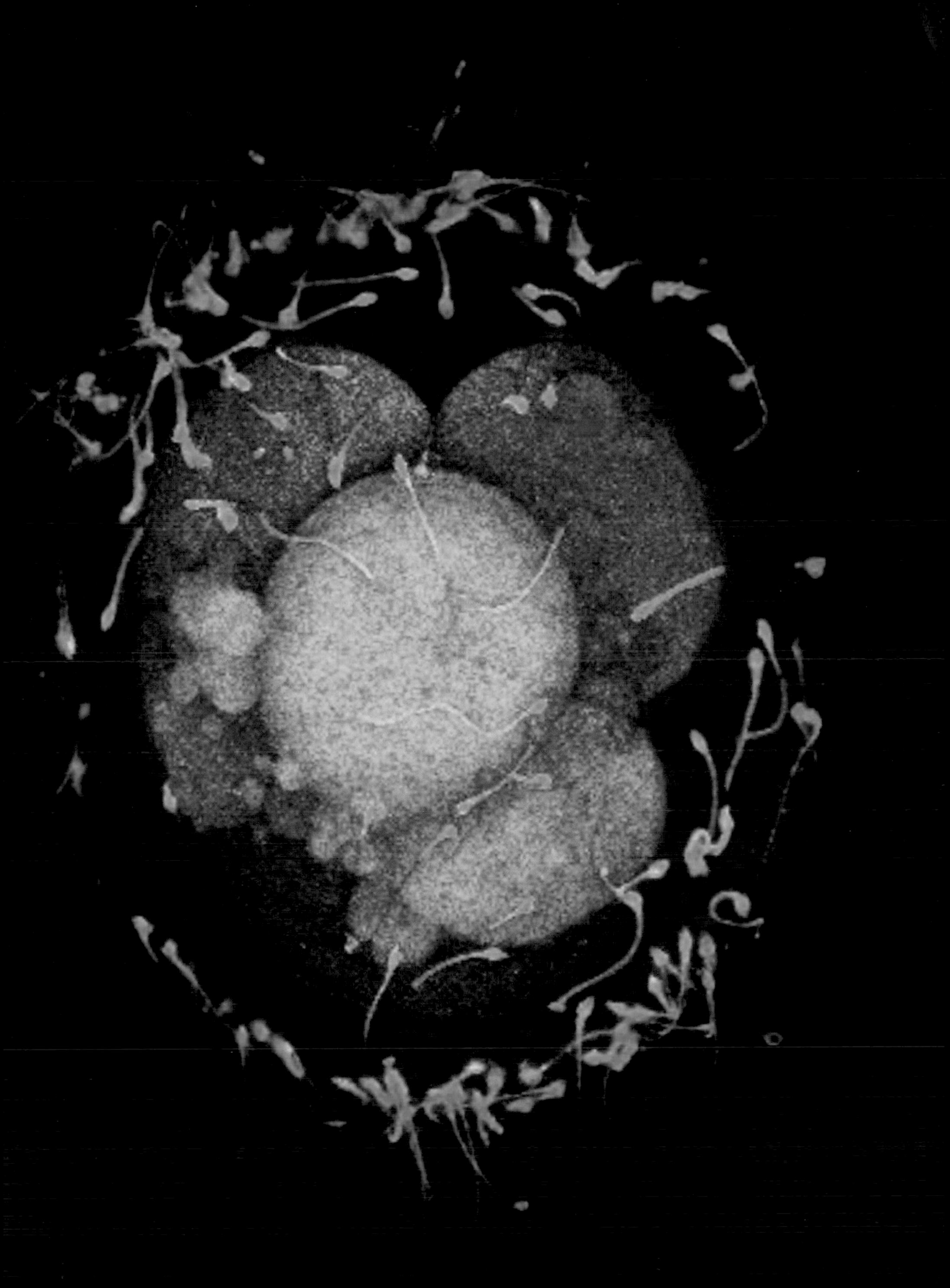

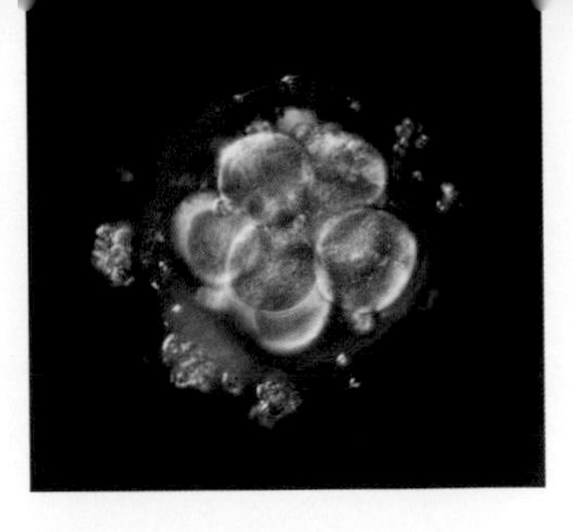

The fertilized egg at the morula stage.

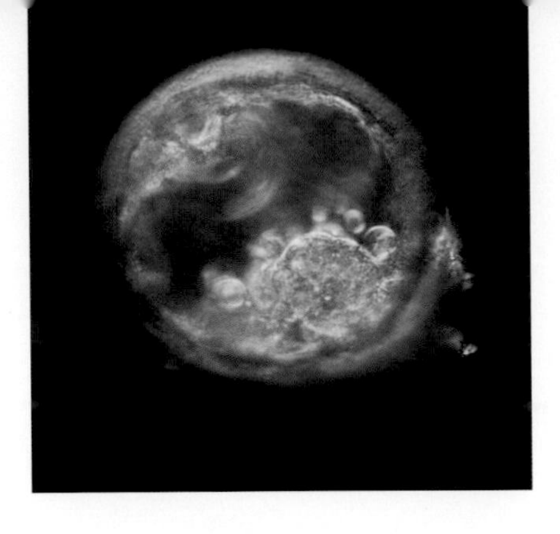

The blastocyst stage, when the cells have begun to form groups.

At this point it is very difficult to distinguish the individual cells, and in the fourth day the fertilized egg (the morula) resembles a mulberry. After another day and night, a hollow becomes visible, after which the morula is referred to as a blastocyst. At this point the first clear division of tasks among the cells can be detected. One group of cells, named the inner cell mass, will develop into the embryo, while others will become the placenta.

The uterine tube narrows into a structure reminiscent of the lock system in a maritime canal, consisting of a muscle surrounding the tube. It has been tightly shut until now, but on the fourth or fifth day after ovulation the lock suddenly opens to allow the blastocyst to pass through. The relaxation of this ring-shaped muscle and its consequent ability to open and close are regulated primarily by the hormone progesterone, which is now being produced by the corpus luteum that was formed by the ovarian follicle after ovulation.

The journey through the narrow part of the Fallopian tube lasts only a few hours. But the blastocyst has to work its way through the folds of the mucous membrane, and it is a very tight fit indeed. Still, the job is a matter of life and death. Should the blastocyst get stuck here, the embryo may implant on the wall of the tube instead of in the uterus, resulting in an ectopic pregnancy. Then the embryo will never be able to develop normally; in some cases the woman will miscarry and in others, she may need medicine or surgery to end the pregnancy.

Four days after fertilization the clump of cells has reached the morula stage and will soon contain twenty-five to thirty cells. Twenty-four hours later it has developed into a blastocyst with seventy to one hundred cells. On their way to the uterus, the cells split into two groups. One will become the embryo, and the other will form the placenta. The picture to the right shows the embryo cells in the lower right region of the blastocyst.

> The fertilized egg is pushed gently forward by the tiny cilia covering the mucous membrane of the Fallopian tube, all swaying in the same direction. This movement is reinforced by rhythmic contractions in the tube's muscles.

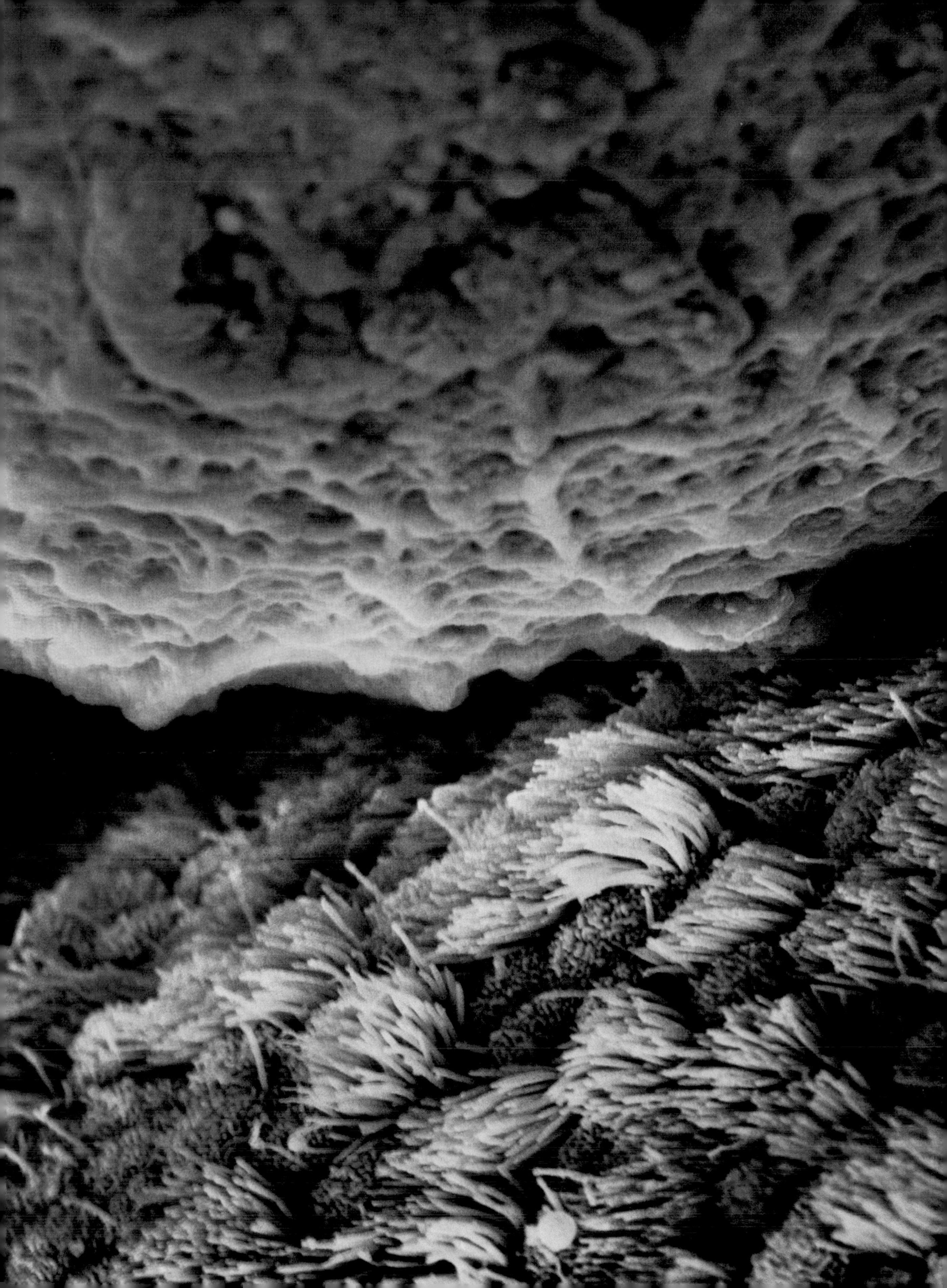

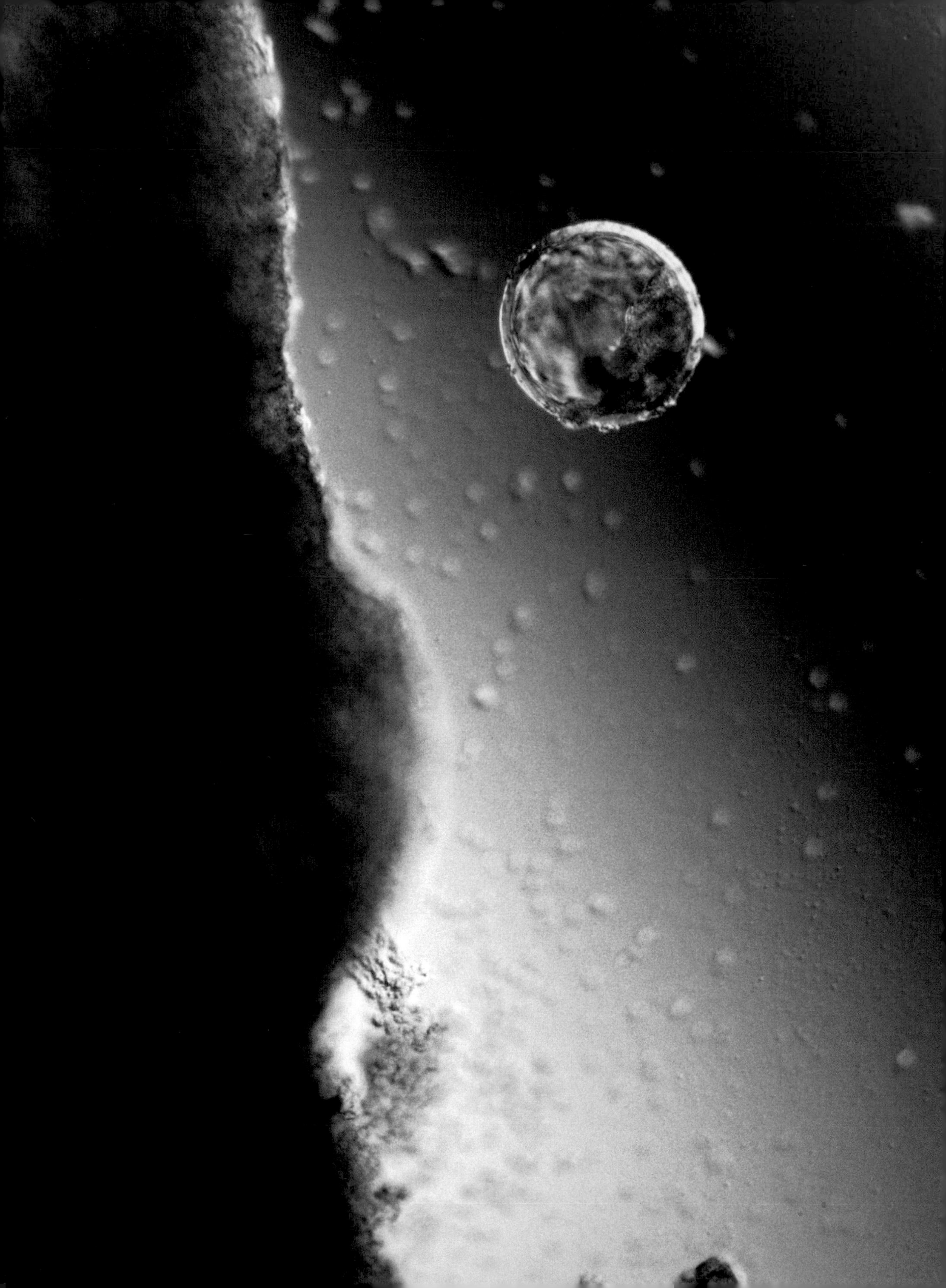

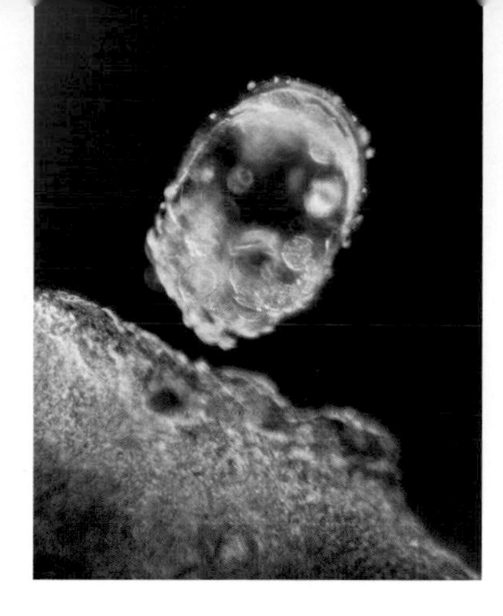

The blastocyst tumbles about.

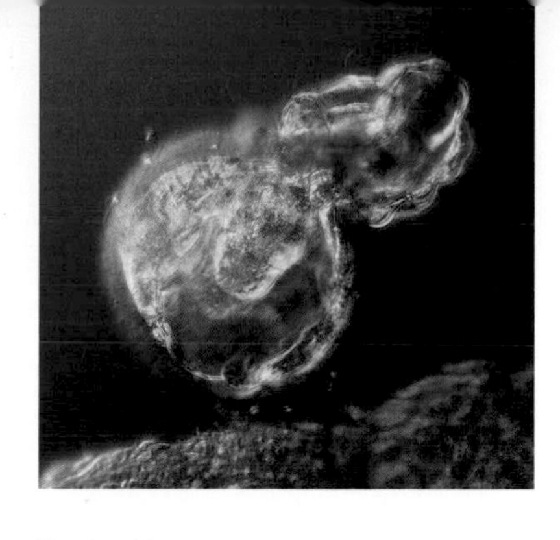

The hatching process.

A soft landing in the uterus

After its tight squeeze through the uterine tube, the blastocyst arrives in the spacious womb. The uterus is ready and waiting. Its mucous lining has had nearly a week to develop since ovulation, while the egg was being released from the follicle, fertilized in the Fallopian tube, and transported to the womb. The embryo chooses its landing place with care. Researchers believe that it sends out chemical signals to its surroundings, which respond to its invitation by creating just the right conditions for it to grow and thrive.

Just before touching down in the mucous membrane, the blastocyst contracts and expands at least three or four times. Together, the embryo and placenta shed their enveloping capsule in a kind of hatching process. If the little embryo is healthy and viable, it will hatch successfully, and the transparent, empty shell sails off and dissolves.

While the outside of the capsule was relatively smooth and hard, the new surface of the egg is more rippled, and stickier. It is as if the entire embryo has been dipped in sugar solution. Shoots of sugarlike molecules reach out for the surface of the uterine mucous membrane, which, in its turn, has similar sugarlike molecules into which the little shoots fit.

The moment of first contact between the embryo and the uterine lining is critical. At this one instant, for implantation to succeed, many factors must come together precisely. It is the embryonic cells, and not the placenta cells, that initiate attachment to the mucous membrane of the uterus, known as implantation. Usually, the embryo implants in the upper uterus.

< The blastocyst prepares to touch down in the uterus, but first it must shed its enveloping shell.

During the hatching process, the blastocyst tumbles around inside the uterus. Now and then it bangs into the soft mucous membrane lining.

(overleaf) Once the shell has broken, the embryo expands extremely rapidly. This friendly intruder gradually settles into the mucous membrane of the uterus.

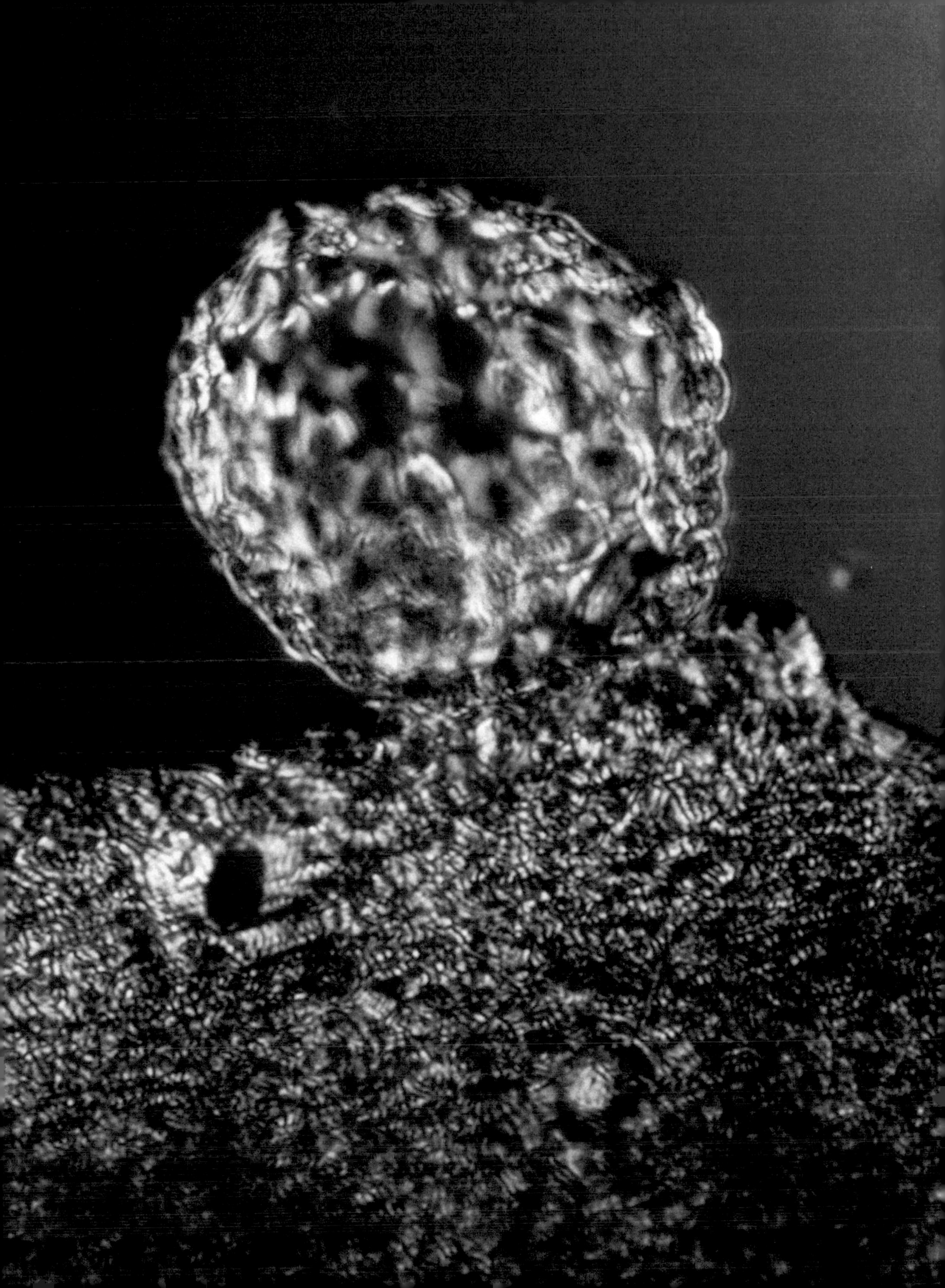

When nature needs help

Although most couples are capable of having one or more children, infertility is a large and growing problem. Estimates indicate that every seventh or eighth couple has more or less serious difficulty becoming pregnant. There are many explanations for involuntary childlessness. Some are well known and clearly defined, while others are more diffuse and ambiguous.

A sperm test will usually reveal if the man is the source of the problem. If the number of sperm in an ejaculation is below five or ten million, and if some of them have poor mobility, there is good cause to suspect male infertility.

Finding out whether the woman is infertile requires more extensive testing. Ultrasound can give the doctor a good idea about the state of the various organs in her pelvic area. It can reveal possible ovarian cysts, or possible fibroids in the uterine lining, and show whether an infection might have led to the Fallopian tubes becoming swollen and blocked. It can show whether the uterus appears normal. A contrast medium is injected through the vagina and the cervix, making it possible to examine via X-ray or ultrasound the inside of the uterus, the thickness of the mucous membranes, and the Fallopian tubes. The test can reveal whether the Fallopian tubes are open. Some hormonal tests are also important, for example to check the anti-Mullerian hormone (AMH). AMH levels give an indication of ovarian reserve and suggest how likely conception would be.

If mechanical and/or hormonal problems are not treated successfully either with surgery or hormone injections, many women turn to in vitro fertilization (IVF) treatments.

IVF treatment begins by giving the woman hormone injections to stimulate the production of mature eggs and prevent premature ovulation. Either a short or a long treatment protocol may be used. In the long protocol, the woman's own hormone production is switched off first, using drugs that act on the pituitary gland. These suppressant drugs are taken in the form of injections for about two weeks before the stimulation phase begins.

The man has provided his sperm sample. Now it is time to retrieve the woman's eggs. If all goes well, they will be fertilized within a few hours.

Just before the time when natural ovulation would have occurred, a thin needle is inserted, with the help of vaginal ultrasound, through the wall of the vagina into an ovary, where it extracts the contents of the follicles. The ideal number of eggs to be harvested is eight to ten, but regulating hormone stimulation so exactly is difficult. The oocytes can usually be quickly

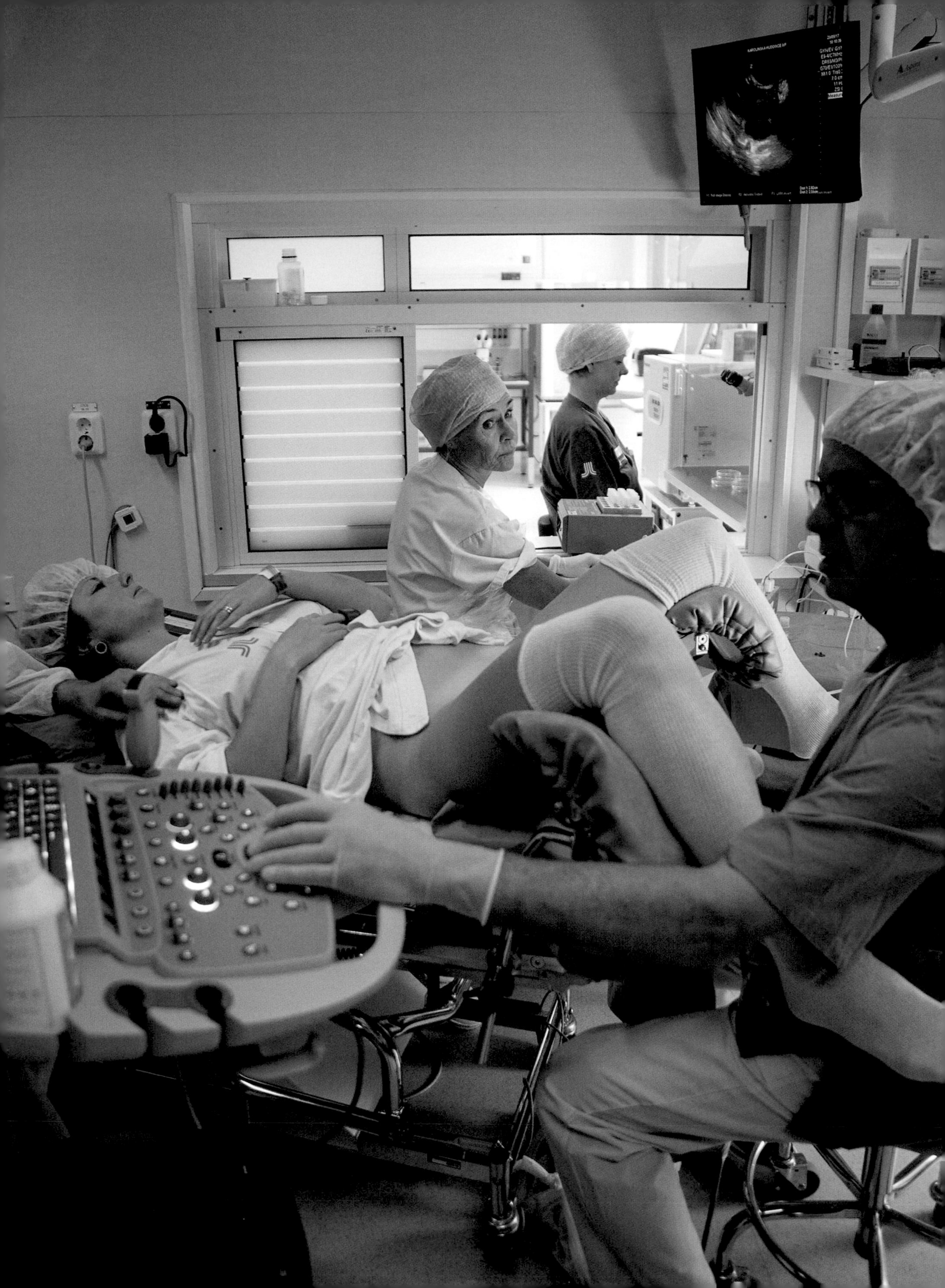

< Guided by ultrasound, the physician collects eggs from one of the ovaries. The fluid with the eggs is placed in a test tube for biomedical analysis. The eggs are examined and counted before being transferred to a culture dish and placed in an incubator to hold them at body temperature. A few hours later, sperm are mixed with the eggs so that fertilization can occur.

After three to five days, a fertilized egg is returned to the woman's uterus.

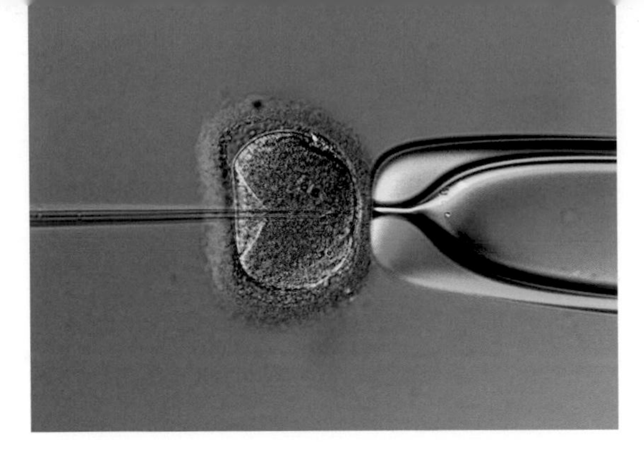

Fertilization using the ICSI method (intracytoplasmic sperm injection).

identified under a microscope and placed in a special nutrient solution in a culture dish. They are held in an incubator at 37.5°C (99.7°F).

After a few hours, fresh sperm, which the man has presented to the laboratory and which are then specially prepared, are introduced into this egg sample. In nature as few as a couple of hundred sperm may reach the egg, but in IVF it is possible to be far more generous. Often many thousand sperm are introduced to the culture dish, in order to maximize the chance of fertilization. After this encounter between sperm and eggs, usually lasting some sixteen to eighteen hours, the culture is examined, and it is easy to see microscopically whether fertilization has taken place. The newly fertilized eggs are transferred to fresh solution, without sperm.

Even men whose testicles produce only a few immature sperm can become fathers with IVF. A doctor can remove sperm from the man's testicles and epididymides under local anesthesia, by needle puncture or minor surgery, then inject a sperm directly into an egg. This technique is called intracytoplasmic sperm injection (ICSI). Even living sperm that are incapable of propelling themselves can fertilize an egg in this way.

When fertilization has taken place, two distinct nuclei appear in the cytoplasm of the egg: one that contains the genetic material from the sperm, and one that contains the genetic material from the egg. A few hours later, still in cell culture, the two nuclei fuse into a unique new genetic code, after which the fertilized egg begins to divide.

After two to three days the egg consists of four to eight cells and can be returned to the woman's body. Currently, fertilized eggs are often cultured for another two or three days, until they reach the blastocyst stage, before being returned to the womb using a thin plastic catheter carefully introduced via the cervical canal. All parties concerned now cross their fingers and hope that the blastocyst is healthy and able to implant in the womb. A week or two later a very sensitive hormonal blood test can be used to determine whether the woman has become pregnant.

In the ICSI method, a pipette is used to carefully puncture the shell of the egg and inject sperm into the egg cytoplasm, where the nucleus is waiting. The egg quickly regains its round shape after the pipette is withdrawn.

> If more than one egg is fertilized and develops into a blastocyst, the extras are frozen. They are stored in long tubes in tanks of liquid nitrogen, in case they are needed in the future.

(overleaf) The fertilized egg has been placed in the woman's uterus and the couple can return home. Sometimes, progesterone is used during the first trimester to help maintain the pregnancy.

HEM

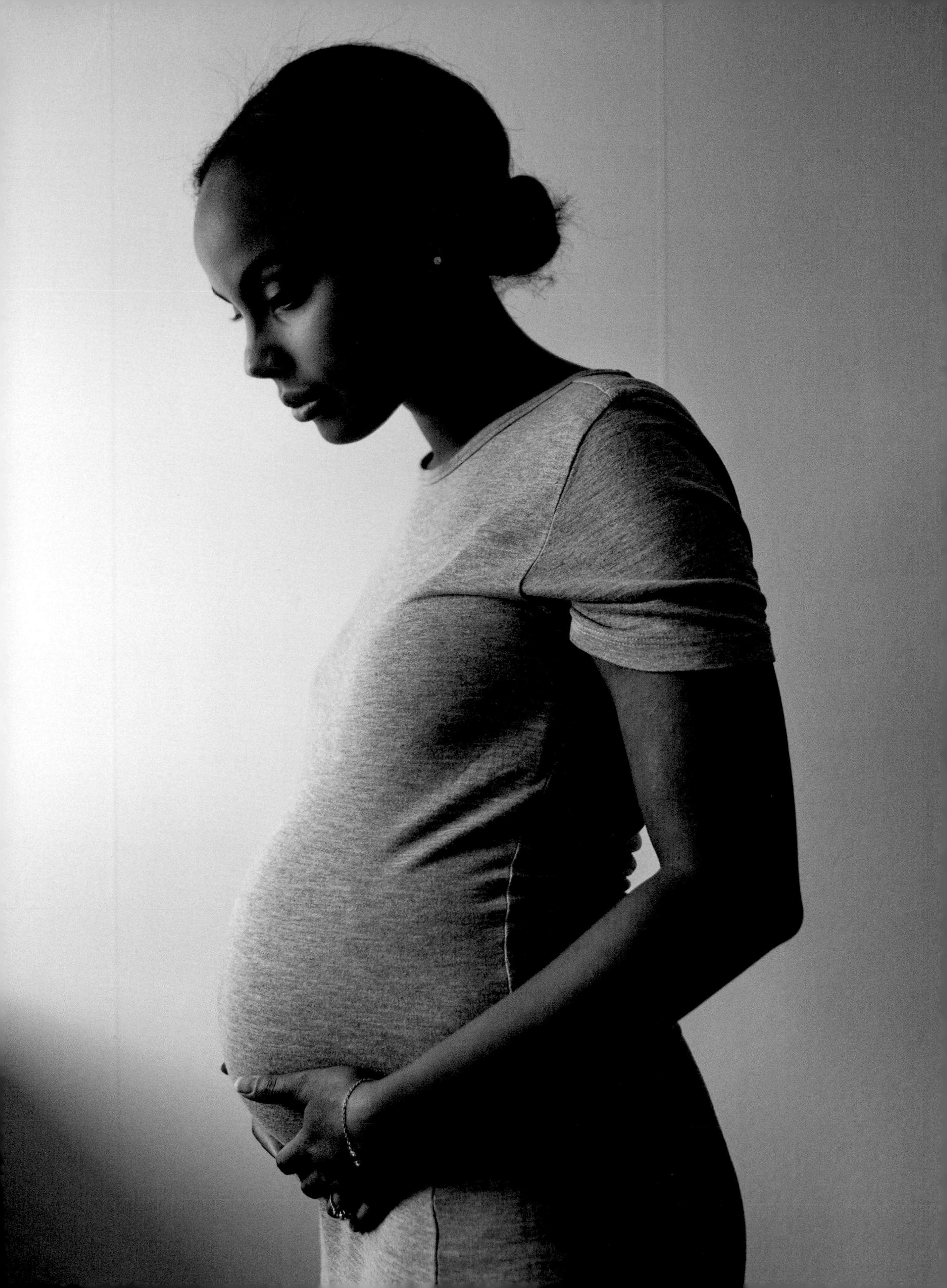

PREGNANCY

Nine months can feel like a long time to wait. But for most expectant parents, pregnancy is an enjoyable and important period of transition: a chance to adjust their routines and get ready to care for their newborn. It is a hopeful time and perhaps a slightly anxious one. The woman feels her body changing as her pregnancy advances. Day by day the embryo and later the fetus grow. On a precise timetable, a miracle unfolds. After 38 weeks a new little person is ready for the world outside the womb.

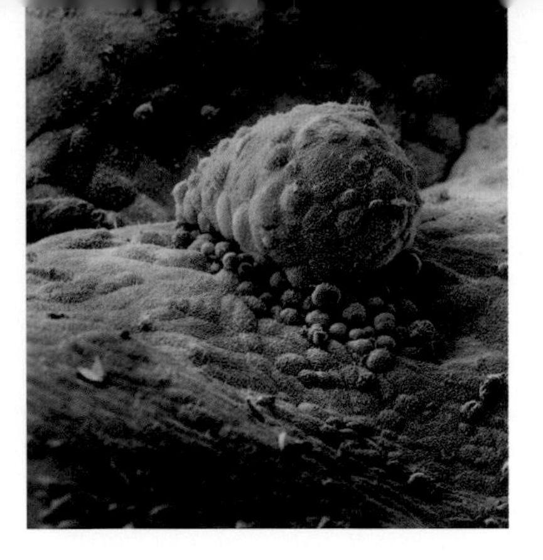

Eight days after fertilization.

Early signs

Right away, the newly implanted embryo begins to solidify its contact with the body of the expectant mother. The placental cells attach themselves firmly to the uterine lining, and some turn into blood vessels that begin to communicate with blood vessels in the wall of the womb. In this way, the mother's blood will provide the embryo with all the nutrition and oxygen it needs and allow the embryo to rid itself of waste products from metabolism.

Some of the cells in the developing placenta give rise to an important hormone, human chorionic gonadotropin (hCG), that signals to the ovaries and the pituitary gland that the woman is pregnant. These signals make it clear that there will be no need for ovulation for quite a while and that the uterine lining should not be expelled—that is: there should be no menstruation. The corpus luteum in the ovary responds to these instructions by creating more progesterone, which reaches the uterus via the circulatory system. Progesterone is the hormone needed by the mucous membrane of the uterus in order to grow and provide the embryo with the environment it needs.

When the embryo implants, the mother may experience light bleeding or spotting. This is normal. Immunologically, the tiny embryo that will develop in the uterus is actually an invader, with a protein composition entirely alien to the mother and therefore subject to rejection. Normally, however, sophisticated systems in the woman's body ensure that the embryo remains and develops without being rejected.

When a woman becomes pregnant, her usual hormonal cycle begins to change, and she may become aware that something is different. Her breasts may become tender, and she may be more sensitive than usual to smells. Mild nausea is not unusual. When her period fails to begin, this will help confirm something she already suspects.

The embryo just after implantation in the uterine lining. Small hillocks in the mucous membrane, called pinopodes, are believed to function as landing beacons.

> As early as about ten days after fertilization, the level of progesterone in the blood rises dramatically.

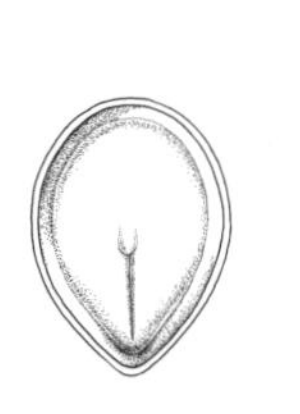

17 days approx.

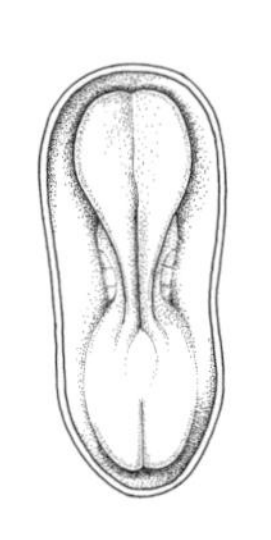

20 days approx.

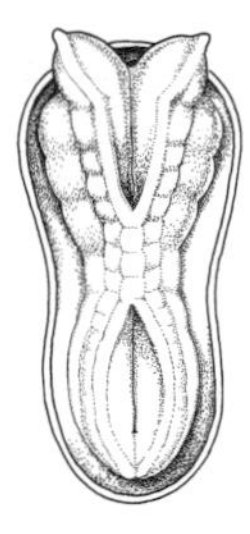

22 days approx.

Rapid, daily changes

The little clump of cells has been attached to the lining of the uterus since about eight days after fertilization. At this point the inner cell mass, the child-to-be, consists of several hundred cells, all of which contain the same genetic code. But no cell can express the entire code at once, and only part of it is expressed in each cell. As the embryo develops, the potential of each individual cell is radically restricted. At a very early stage, the embryonic cells are divided into three layers, called the germ layers. The cells in the outermost germ layer will go on to become the brain, spinal cord, and nerves. Over the course of a few days, this layer thickens around the midline of the body, forming two lengthwise folds. Between these folds an indentation deepens, then quickly closes, making a tube. At one end the primitive brain begins to form like a little bubble. Nerve fibers begin to protrude from the brain stem, and the spinal cord starts to form. Other cells in the outermost germ layer will go on to form the skin, hair, sebaceous glands, and sweat glands.

Cells from the middle germ layer will develop into the skeleton, the heart muscle wall, and the other muscles. They will also become blood vessels, lymphatic vessels, and blood corpuscles that, together with the heart, will form the circulatory system. The ovaries and testicles, as well as the kidneys, will also come from this layer. The inner germ layer will give rise to the intestinal system and the urinary tract. This layer will also provide the mucous membranes of the entire body as well as the lungs.

This early stage is an extremely sensitive one, as even the slightest deviation from the program may result in damage and deformities that the child-to-be will carry throughout life. During this time both the external and the internal environment are incredibly important. Not only the individual organs but also the communication between them has to develop perfectly, down to the minutest detail. If a serious defect arises, natural protective mechanisms go to work, and the woman has a miscarriage. Although miscarriages may have other causes, genetic abnormalities are the most common. Nearly one pregnancy in five comes to an end at this stage, because either the egg or the sperm lacked the optimum genetic prerequisites.

COUNTING THE WEEKS

Pregnancy length is counted in weeks, beginning from the first day of the woman's last period. The actual moment of fertilization is usually about two weeks after the last menstrual period. Therefore the embryo and fetus are always about two weeks younger than the pregnancy week suggests. Unless otherwise noted, the "weeks" in this book mean pregnancy weeks (also called gestational age). When doctors and midwives give a pregnancy week count, they always count completed weeks. Thus, a woman at "week 9, day 3" has already begun her tenth week of pregnancy.

When week 5 begins, the embryo changes rapidly. Within just a few days it is transformed from a clump of cells into an oblong body, with a head and a tail beginning to take shape around the neural tube.

> The embryo is still just a few millimeters long. Its backbone is curved, with the neural tube open at both ends of the body. The outer germ layer, the skin of the embryo, now encloses the torso.

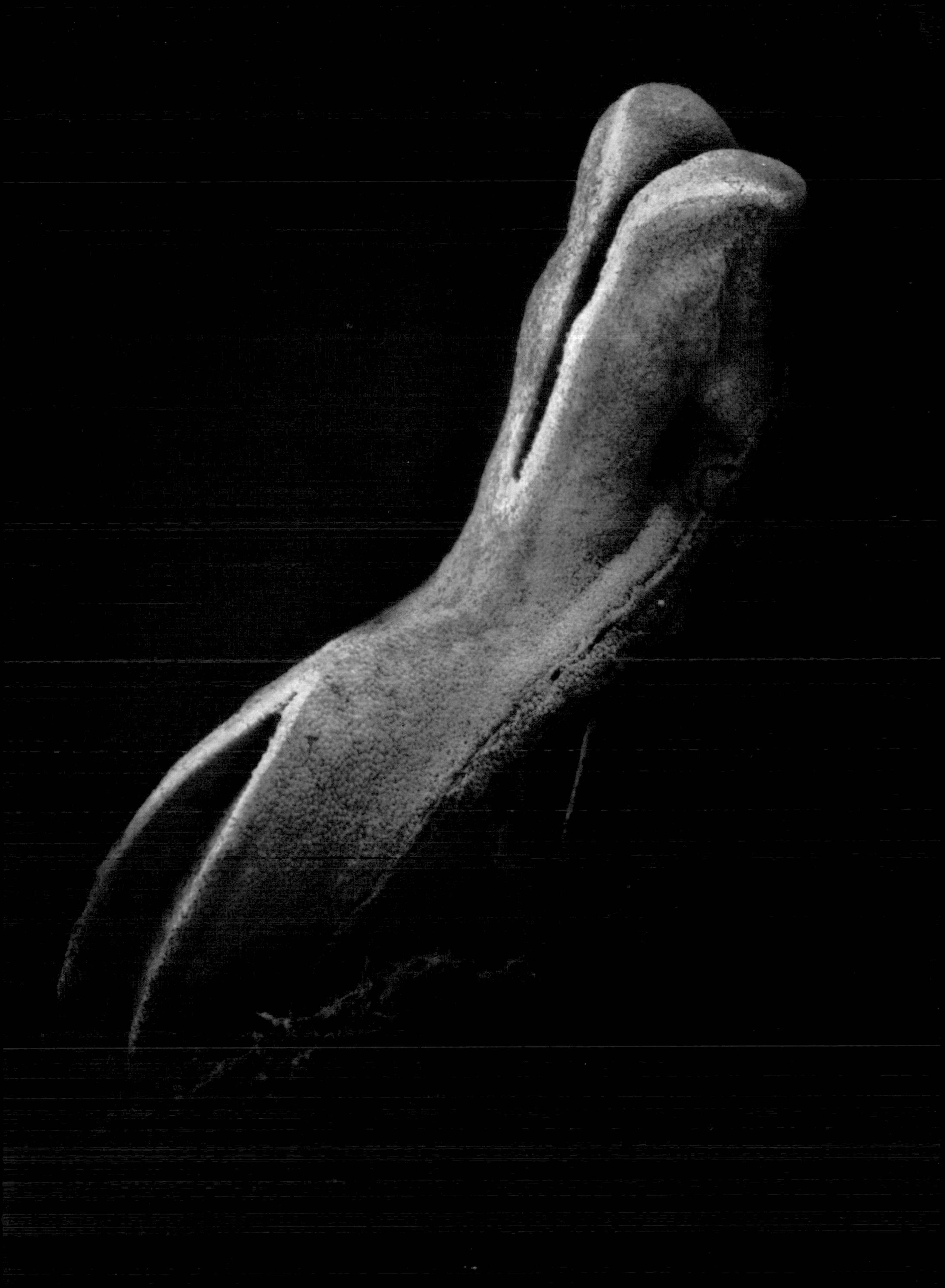

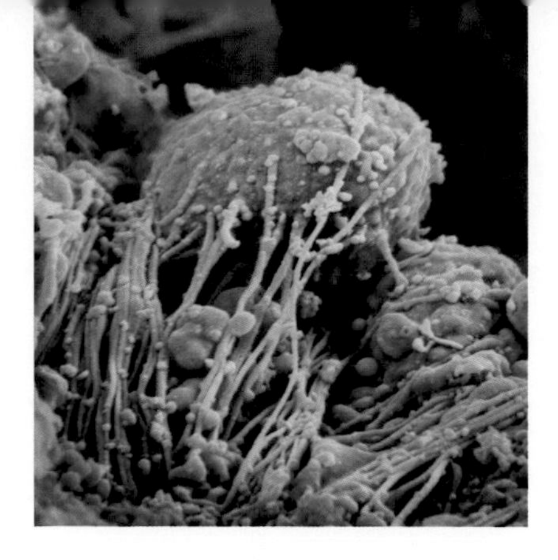

Brain cells from a 23-day-old embryo.

The early brain

Just a few weeks after fertilization, primitive nerve cells, initially almost round, are visible in a human embryo. Gradually, long runners known as axons develop from the bodies of these nerve cells. They establish contact either with other nerve cells or with an end station, such as a muscle, to which they will report information. Each nerve cell transports information mainly via electrical impulses, with a chemical transfer at the points of contact.

One of the substances responsible for this chemical transfer is noradrenaline, which rapidly relays information to the muscles and elsewhere. Dopamine and adrenaline are similar chemical transmitters. Still another chemical, acetylcholine, transfers information to the stomach and intestines but starts to work later in fetal life. Although the acetylcholine transport system is far less rapid, it is no less effective.

The dynamic and stringently programmed development of the brain is essential for the growth of the body, as well as for the ability of the arms and legs gradually to begin to move in accordance with the normal pattern. The nerve stems, parallel bundles of axons, develop early and may be compared with broadband telecommunication, transporting complex information throughout the body at great speed.

After some time the different parts of the brain become more clearly differentiated and take on their separate tasks. Some parts of the brain receive only sensory impressions from the body, such as sensation and pain, while others are responsible for vision and hearing, and still others govern movements, which begin awkwardly and gradually become better coordinated and more purpose-oriented.

Little outgrowths (axons and dendrites) have begun to protrude and are striving to make contact with cells in their proximity. A nerve cell receives a bunch of dendrites, delivering information.

> The 22-day-old embryo does not yet have a face. At first the brain is wide open and unprotected, but soon it is covered by the skull bones; these are joined only loosely to leave space for the brain to grow.

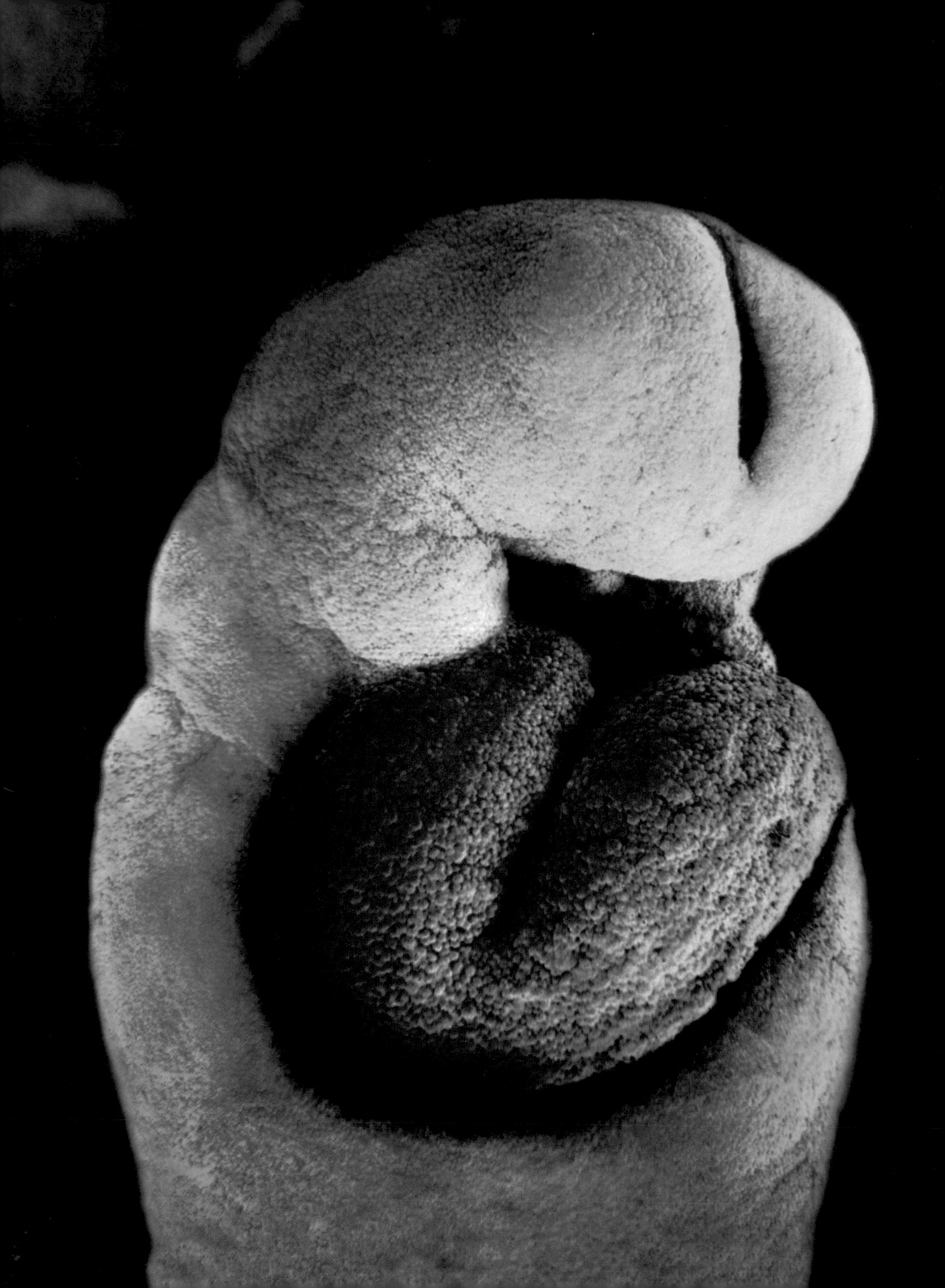

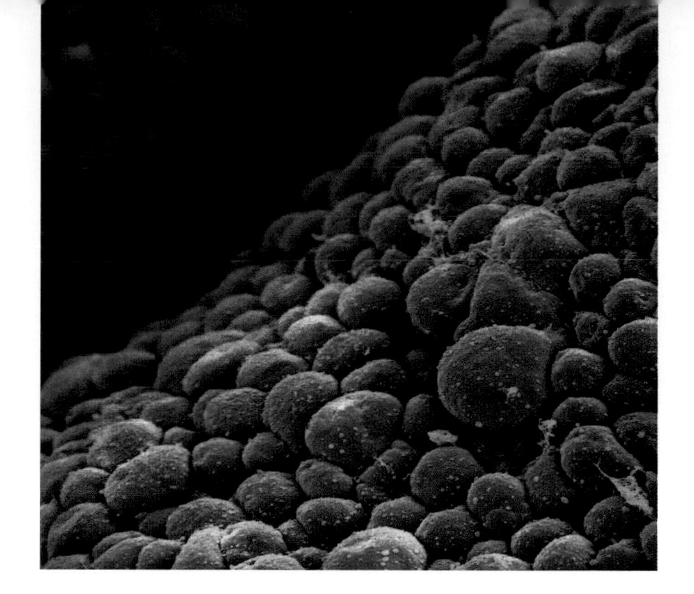

The heart muscle cells of the embryo in close-up.

The first heartbeats

The heart begins to develop when the embryo is still but a cluster of cells, and as early as its twenty-second day the newly formed heart muscle cells contract, and the heart beats for the very first time. The mission of the heart, central to the development of the embryo, is the circulation of blood, which distributes nutrients and oxygen to all the tiny developing organs.

At this stage the heart already has two chambers (ventricles) and is so large that it almost seems to be outside the rest of the body. The right-hand chamber takes up the blood from the other body organs, and the left-hand one releases freshly oxygenated blood to the body. After birth, blood from the body will return to the right-hand chamber, which will pump it into the lungs to pick up oxygen. Then blood will flow into the left-hand chamber, where it will be pumped into the aorta, the large blood vessel that distributes blood throughout the body. In the fetus, however, the lungs are collapsed. The placenta oxygenates the blood, which returns to the heart to be directed to the rest of the body. At birth the lungs will expand and take over from the placenta.

The heartbeat of the embryo is very rapid, nearly twice that of the mother, and can easily be heard even with very simple listening devices. Heart rate is one of the most reliable ways of knowing how the fetus is faring.

When one muscle cell contracts, there is a domino effect on the surrounding ones. No separate heart nerves yet govern the heartbeat.

< When the embryo is 24 days old, the primitive heart accounts for much of the body. For two days it has been ticking fast and rhythmically.

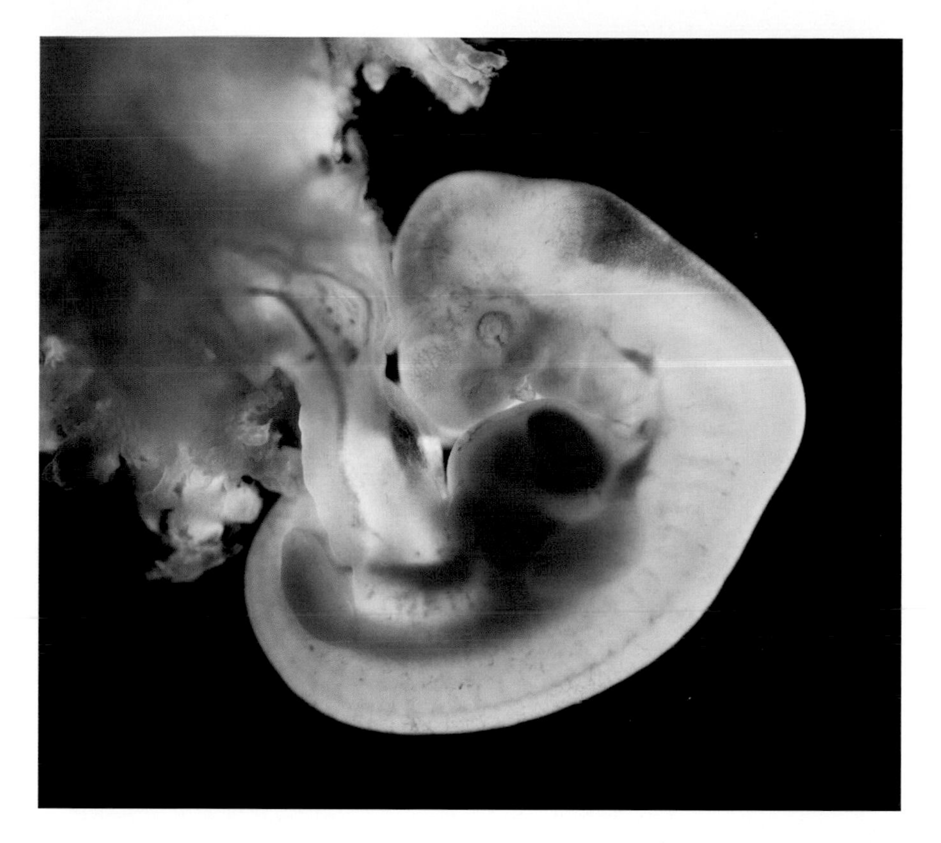

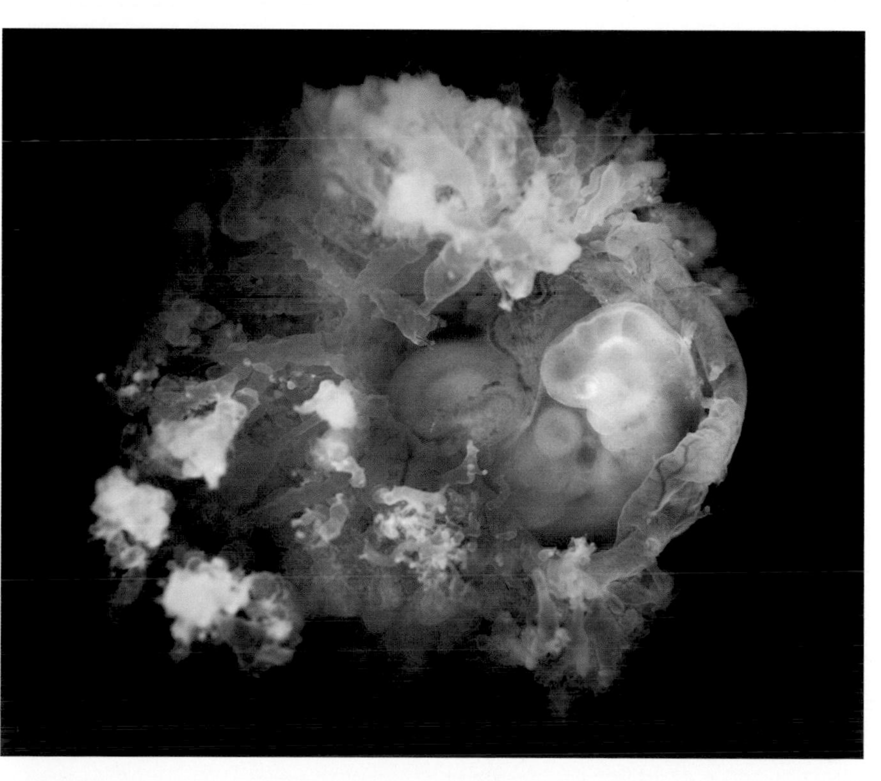

< A four-week-old embryo (pregnancy week 6), seen from the back with the primitive brain to the right. The head end is growing quickest.

> The placenta is quick to entwine itself with the uterine lining and blood vessels. The connection is like a busy freight depot, with incoming nutrients from the mother's blood passing metabolic waste products on their way out. Five weeks after fertilization (pregnancy week 7), the heart and liver are disproportionately large and the hands and feet are only tiny buds.

> The embryo floats comfortably in the amniotic sac, cushioned by the amniotic fluid. The round ball under the head is the heart. To the left of the heart, the primitive vertebrae and the tail are visible.

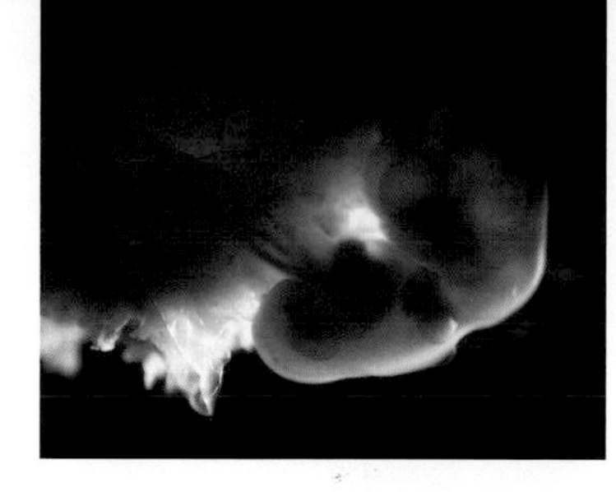

Week 8.

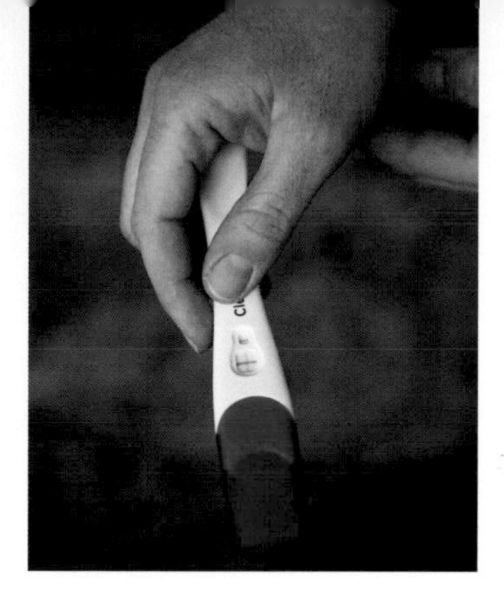

Blue line = pregnant.

Confirmed!

Today there are simple, sensitive tests a woman can use to verify whether she is pregnant as early as her first missed period and sometimes even earlier. Pregnancy tests work thanks to a hormone that develops very quickly after fertilization. This hormone, human chorionic gonadotropin (hCG), enters the woman's circulatory system and can therefore be traced in her urine after a few days. Some modern pregnancy tests can even show how far along the pregnancy is.

When will the baby be born? Although there is no precise length of pregnancy, a woman will usually be given an estimated date, for both medical and psychological reasons. This date is based on an "ideal" pregnancy, lasting forty weeks from the last menstrual period. One traditional method of estimating the day the baby will be born is to count from the first day of the last period, subtract three months, and add seven days. This will work only if the woman has had regular periods at twenty-eight-day intervals. If the intervals are longer or are irregular, calculating the due date will be far more difficult. In recent years ultrasound has come to be considered the most reliable way of dating the age of the fetus and the delivery date. Many women have their first ultrasound exam at around week 18, but an ultrasound to confirm the due date may also be done earlier.

Most women who become pregnant are thrilled, and the idea of a baby is a very welcome one, but other women must, for one reason or another, make the difficult choice of terminating the pregnancy. The number of abortions varies from country to country and depends on the legality of abortion and the availability of contraceptives. In the United States and the Nordic countries a woman has the right to terminate her pregnancy in its early stages. In other countries, the laws prohibit abortions, and a woman must undergo the procedure illegally, which has serious risks. A poorly performed abortion can result in future sterility because of infection. Globally, one of the most common causes of death in women between the ages of fifteen and forty is complications after an illegal abortion.

Six weeks after fertilization (pregnancy week 8), the embryo is about 10 millimeters (0.39 inch) long. The cells are bursting with life. The heart is beating and blood is pumping through the vessels in the umbilical cord.

< It is easy to tell whether a pregnancy has begun. All it takes is a plastic stick with a color indicator. The stick is dipped in a urine sample; if the color changes, the woman is very probably pregnant.

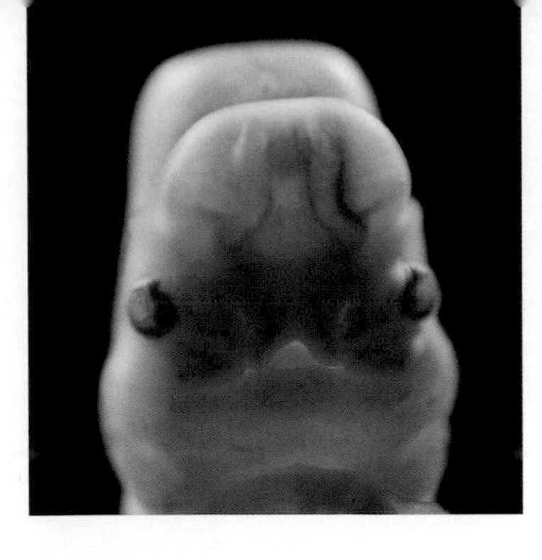

Portrait of a 30-day-old embryo.

The body and face take shape

The embryo is now five weeks old and well past the stage when it looks like a formless clump of cells, but the tiny body is still quite transparent. The head is dominant, the upper body is much larger than the lower body, and the arms and legs have begun to develop. As early as week 5, a clear line develops down each side of the body, from what will be the shoulders to what will be the hips. Just over a week later, two little skin-covered buds appear at either end of each line. These buds are raised but flattish, reminiscent of the mitts of a seal. Soon they develop an edge, which then begins to protrude. At the top end this protrusion signals the connective tissue to prepare the hands and upper and lower arms, and at the other end, somewhat later, the feet, thighs, and lower legs. The hands begin to develop much more quickly than the feet, and maintain their head start for some time: the tiny baby will learn to grasp long before learning to walk. The developmental timeframe is precisely programmed and varies hardly at all from one individual to another, although genetic variation between individuals can otherwise be quite large.

The face, too, is now taking shape. It begins to form when five projections, called processes, gradually develop under the thin layer of skin and then meet. The first process protrudes between the eyes, ending in a bay on either side, the primitive nostrils. Gradually this process becomes the nose and the center of the upper lip. Two other processes shoot out from the sides of the head, one under each eye, to form the cheeks and the sides of the upper lip, and the last two develop below the mouth, forming the lower lip and chin. From this skeletal framework, the muscles grow that will allow the face to move and adopt expressions. Any little disruption of this development, such as a viral infection lasting a few days, may result in a cleft lip, a defect that often upsets parents terribly but that can be surgically corrected with excellent results.

Now we can see the eyes, nose, and mouth—a developing face. The nostrils have formed above the opening that will be the mouth, but the groove between the nostrils and the edge of the top lip has not yet been effaced. The little hollows that will be the eyes are almost all the way out by the temples.

> The embryo at week 8. What we see highest up is not the head, but the neck. The head is still curled forward, chin tucked to chest. The torso accounts for no more than half the body length.

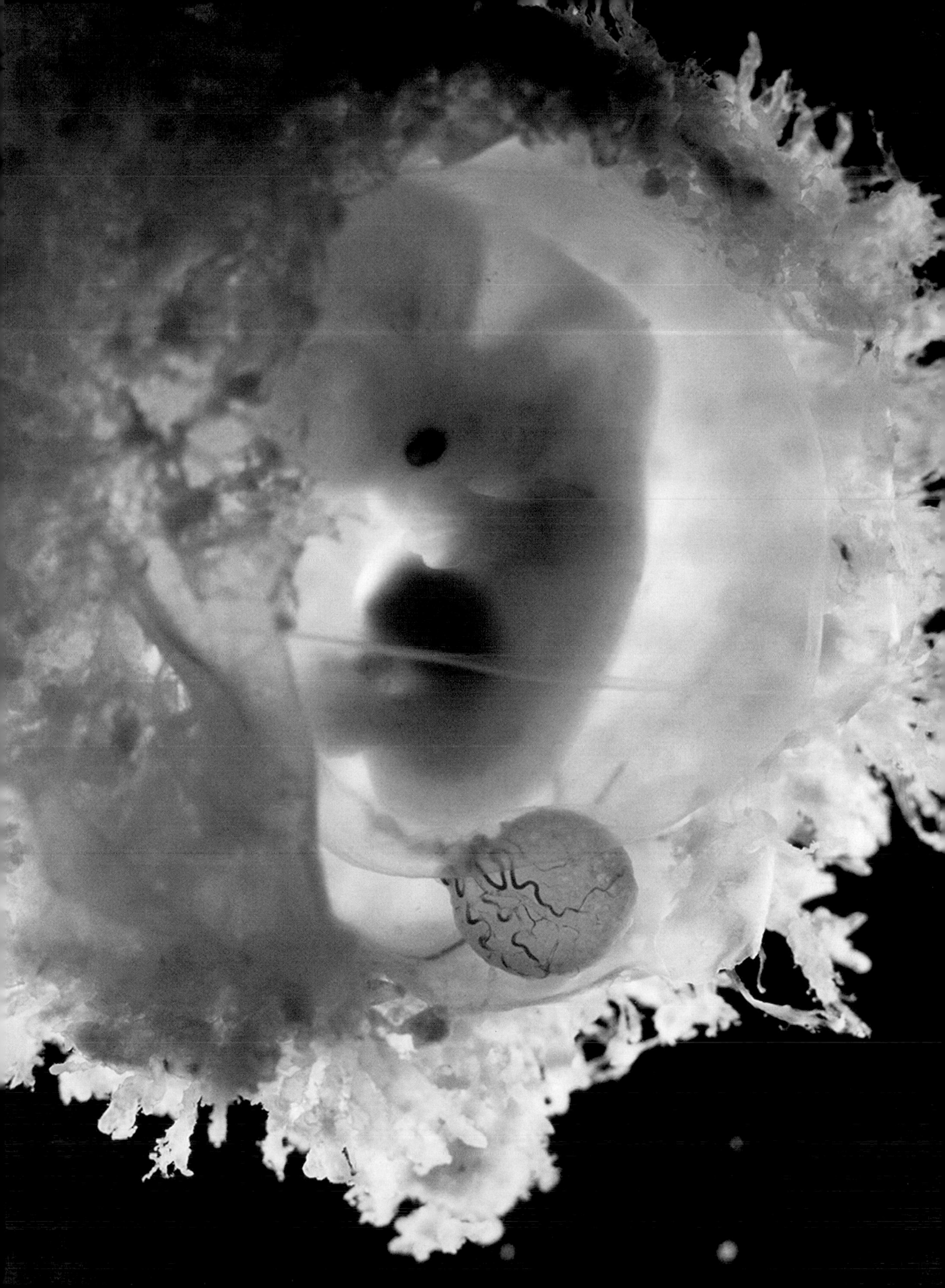

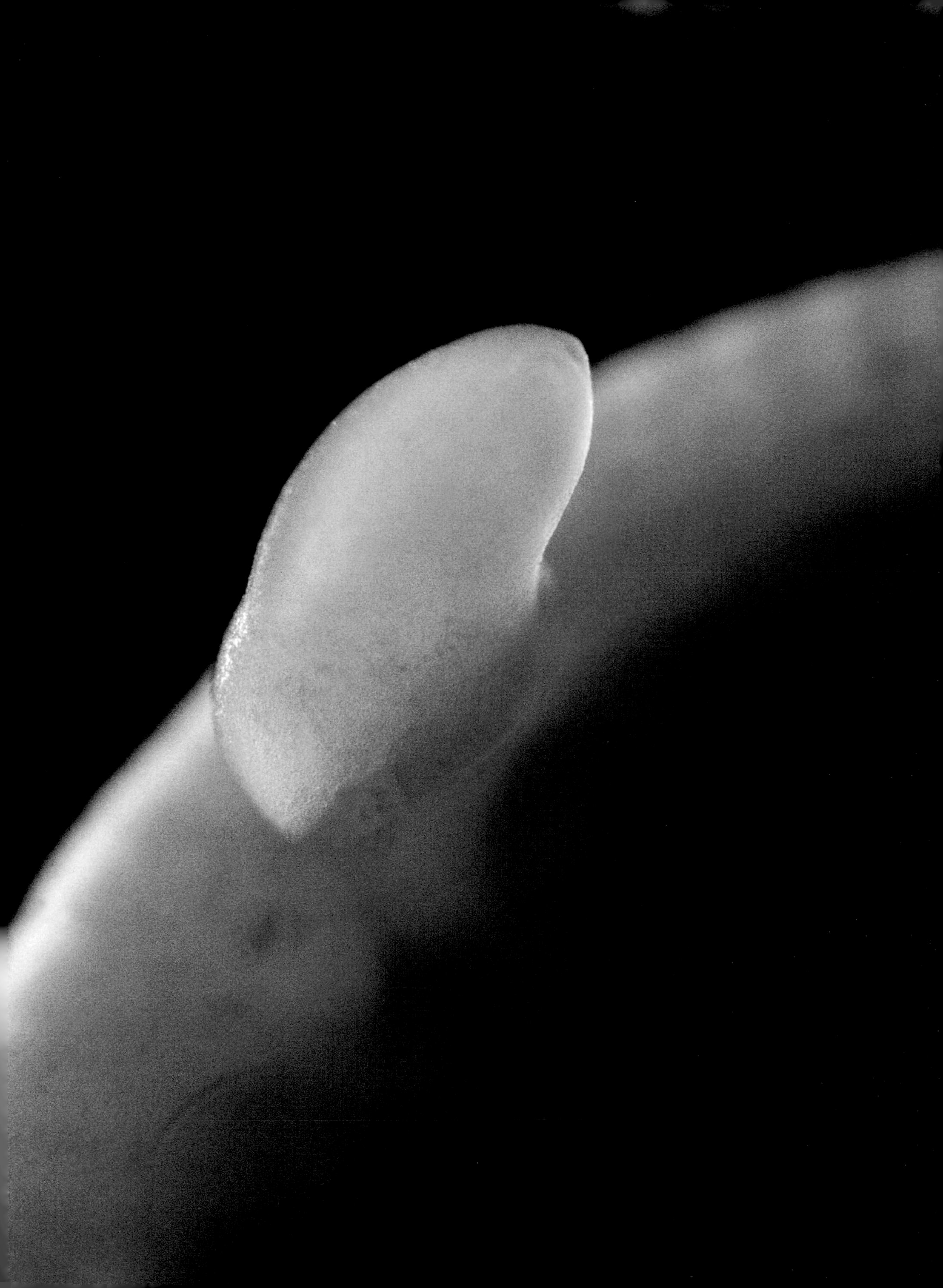

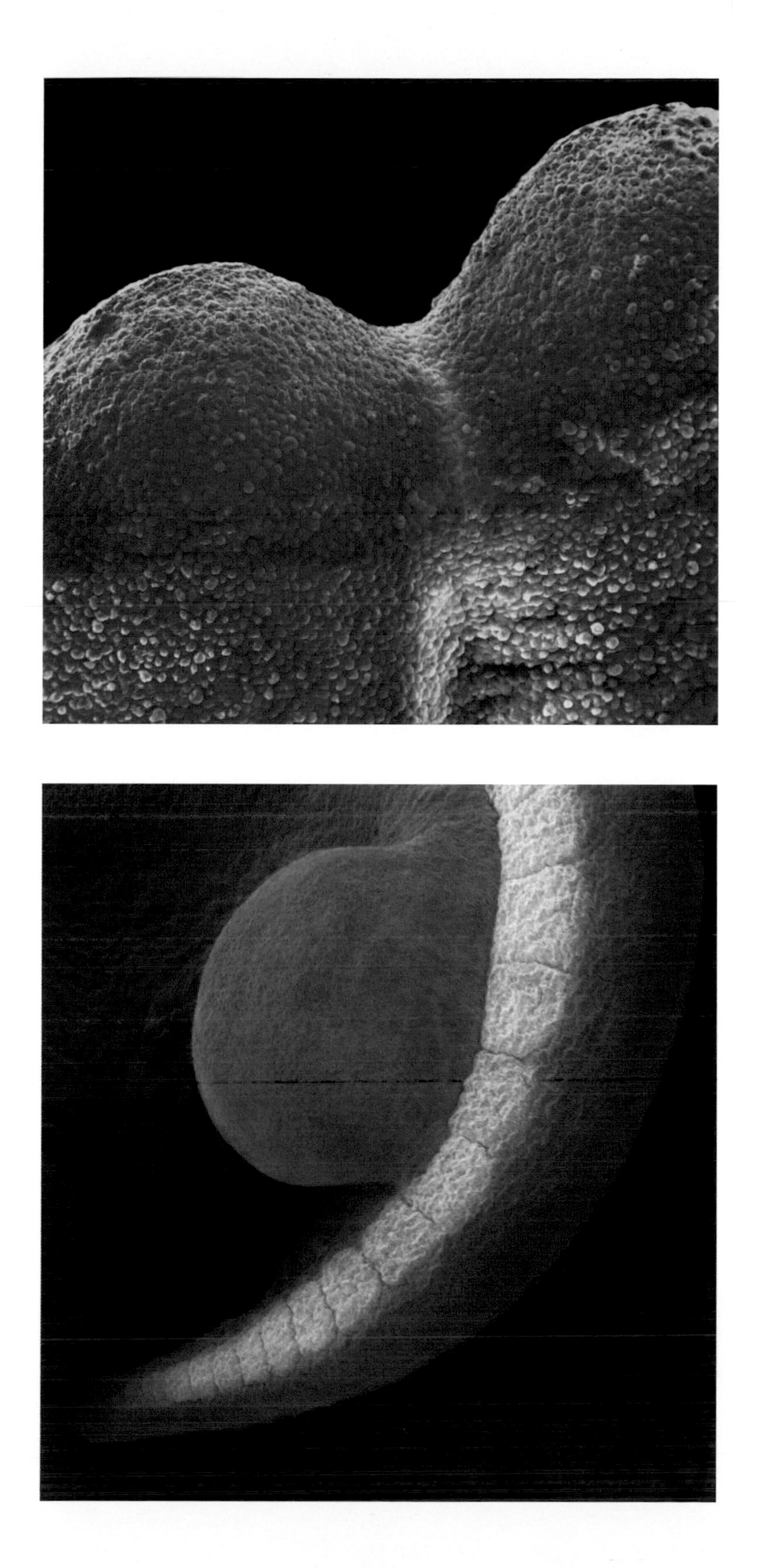

< A budding arm protrudes from the torso as early as week 6.

> Two weeks later there is a suggestion of fingers. The tissue between the fingers regresses.

> In week 6 the primitive feet and legs are no more than buds protruding from the backbone. The vestigial tail vanishes early on, but we do retain a tailbone, a few small, stunted caudal vertebrae, all our lives.

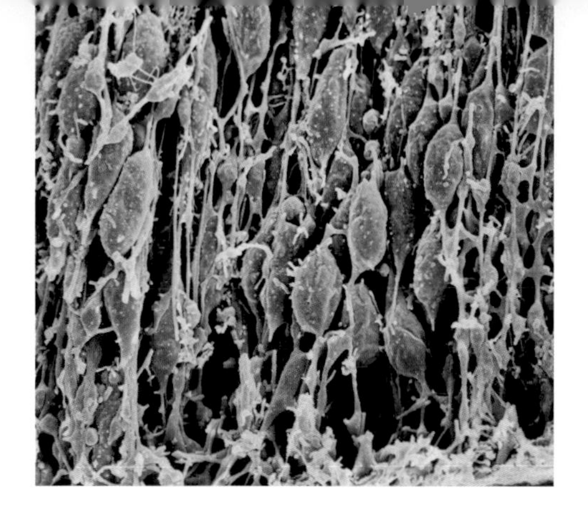

Nerve cells in the brain, week 8.

Signals to and from the spine

Even in the embryonic stage, human beings are clearly vertebrates. On both sides of the trench of nerves along the backbone, skeletal building blocks form from the middle germ layer. These develop into thirty-three or thirty-four vertebrae, although the four caudal vertebrae, left from a past stage of evolution, fuse into a single bone, the coccyx or tailbone. Those closest to the head are called cervical vertebrae.

Between the primitive opening that will become the baby's mouth and the bulging heart sac, seven shoots develop from the vertebral column. At first these shoots look like fish gills. Soon, however, one of them grows into the baby's jaw, while the others become the face and neck. Below these the twelve thoracic vertebrae are formed. The ribs grow out of these vertebrae, arching to shape the baby's chest. Inside this cavity there are already primitive lungs. The vertebrae must not fuse–if they did, the backbone could not bend. Elastic tissue and muscles will hold the vertebrae together and gradually steady the backbone.

From little hollows between the vertebrae, tiny bundles of nerves reach out and spread a delicate web through the body. This network will play two very different and important functions: the brain and the spinal cord will emit signals to all the muscles in the body, instructing them to contract and perform various motions; and information will be returned to the brain via the spinal cord, informing the brain of what the rest of the body is doing.

This signaling system begins to operate fully in pregnancy weeks 8–9, when the embryo is six or seven weeks old. Some of the nerves informing the brain register touch, while others register pressure, temperature, and so on. Other special nerve impulses come to the brain from the eyes and nose and from the mouth and tongue. Thus an entire nerve structure serving our senses is constructed very early in life indeed.

As early as week 7, nerve cells in the brain begin to stretch toward one another. Some have already connected, laying down primitive neural pathways. More and more systems are getting hooked up and starting to talk to one another.

< Week 8. Here we see the whole vertebral column, running down from the neck to where the legs will be. The arm buds extend like little wings. The placenta is much larger than the embryo at this point. Suspended on the left, hanging like a balloon, is the yolk sac, which, among other things, may influence the development of the genitals.

A little human begins to take shape

In weeks 7 and 8, the face, torso, arms, and legs continue to grow. What formerly resembled the embryo of any primitive mammal now begins to look like a miniature human being. The head, hitherto tipped far forward, straightens up. The skull bones are not yet very ossified, so the brain shows right through the thin cartilage. In the forehead area, two big bubblelike structures will become the cerebrum, and behind it three smaller ones will eventually build up other important parts of the brain.

The head is quite huge in relation to the rest of the body, because the growth of the embryo takes place from the head downward. Not until later in life will the body completely catch up, and in a newborn the head still accounts for about one quarter of the body's total length, compared to one eighth in an adult. By week 6, it is usually possible to measure the exact length of the embryo with an ultrasound scan, and to see its rapidly beating heart. The length measurement is given as crown rump length, or CRL: the length of the embryo from head to bottom, in a sitting position.

Many of the organs have already begun to function: the kidneys produce urine, and the stomach produces gastric juice. The embryo has also begun to move: the first "visible" motion is the rapid, steady beating of the heart, but soon small bodily movements show that nerve impulses coming from the brain are instructing muscles to contract. These begin as global motions, affecting the whole body, but gradually specific little movements take place, such as one hand moving while the rest of the body is still. This constant motion is important, stimulating normal growth and development of the muscles and joints.

Even this early in pregnancy, the embryo is extremely lively, in constant motion, sleeping for only brief periods. The nerve fibers that govern motion and receive all the sensory impulses extend into the arms and legs.

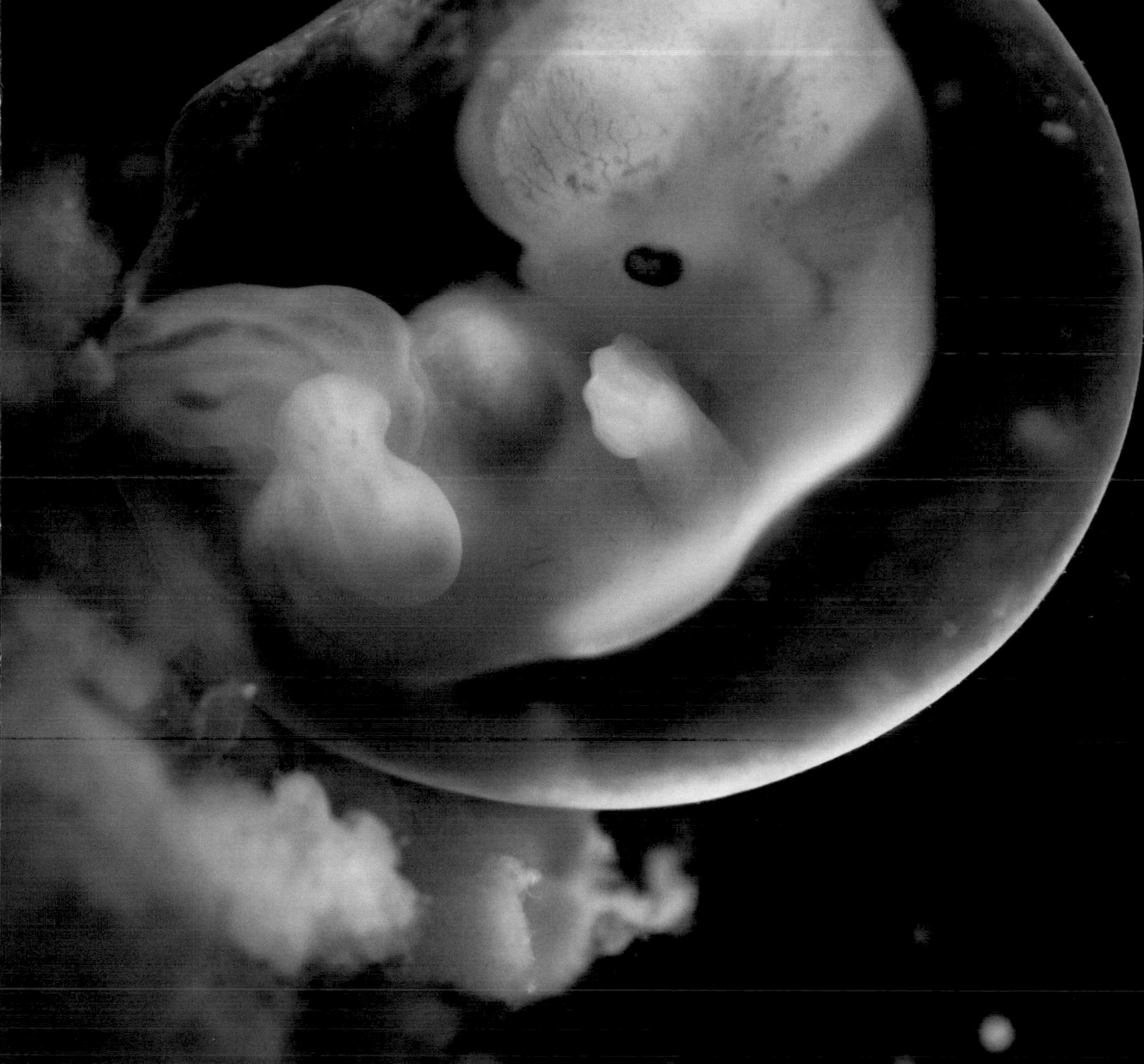

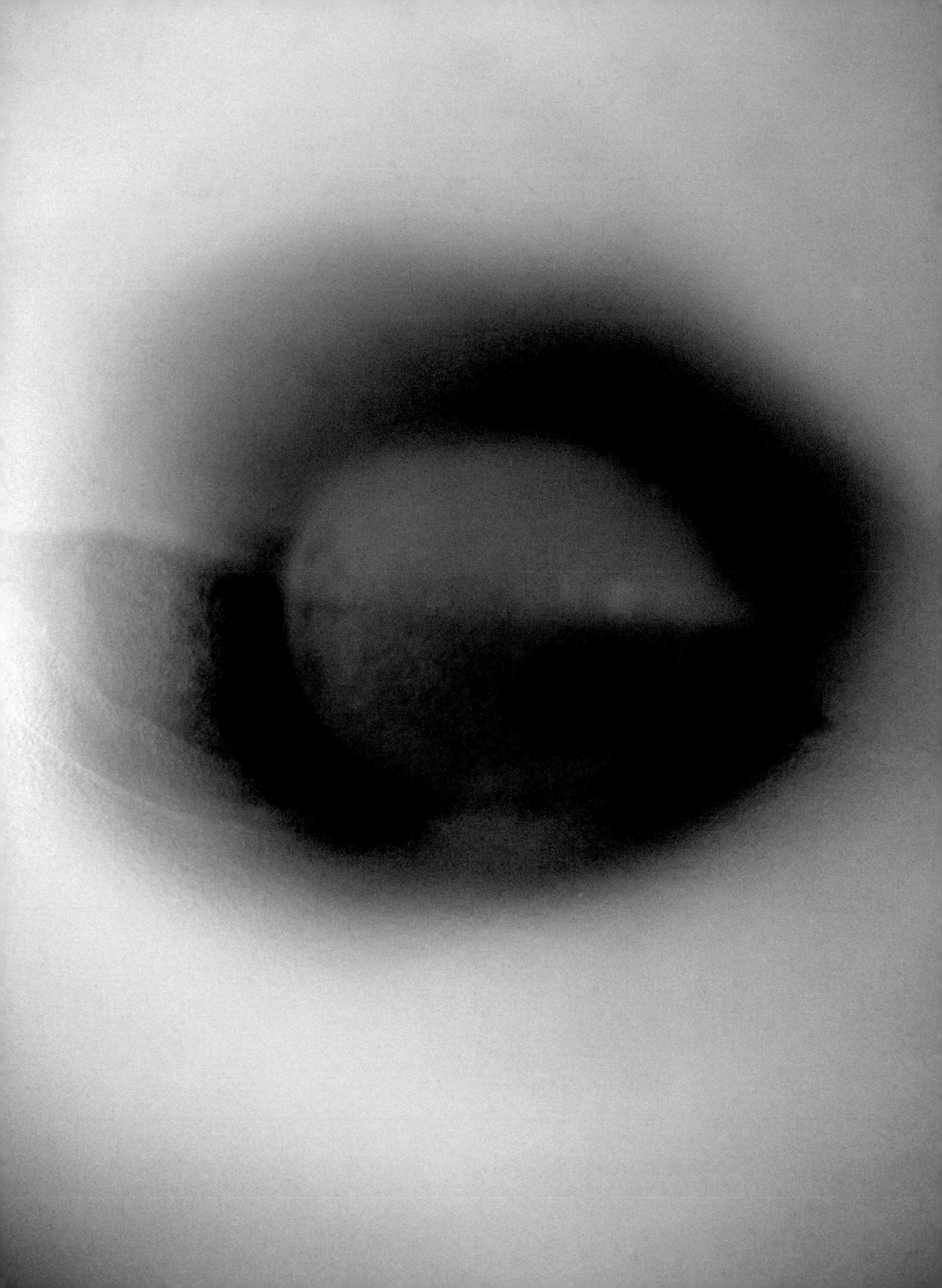

The eye in week 7.

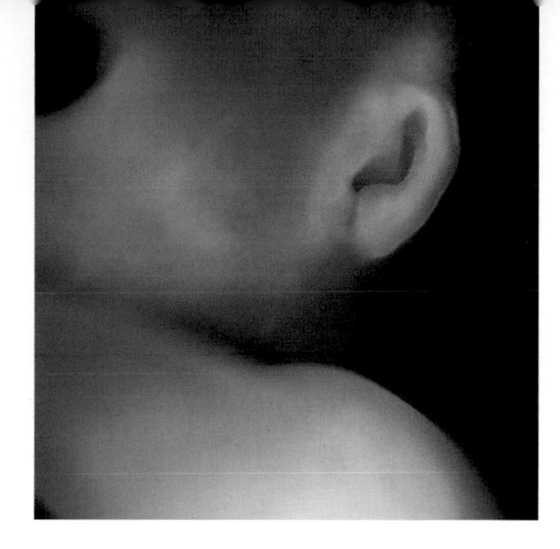

The ear in week 12.

Eyes and ears

The eye is clearly established in the embryonic stage, and soon the lens develops, a spinoff of the skin. In week 10 there are primitive eyelids. The outer ear begins to form in week 8. Its development from a shapeless little bud to a finely chiseled, perfectly shaped organ takes several months.

< In week 12 there are primitive eyelids. These quickly cover the eyes, which will not open again until week 26.

Life in the uterus does not require a developed sense of vision, yet the eyes begin to develop early in the embryonic stage. First the frontal portion of the brain sends out a hollow shoot toward the skin on either side of what is going to be the face. The end of this stalk is a puffed little bubble, the beginning of the eye. When this bubble reaches the inside of the skin, it curves inward, like a cup. The bottom of the cup becomes the bottom of the eye, and the cover of skin is transformed into the retina. In the cavity of the cup a lens gradually forms, as well as a cornea. In front of the lens the iris takes shape, growing from the edges toward the middle. Last of all, two flaps of skin fold down to become eyelids. The eye is complete.

The development of the inner ear also begins early. It resembles that of the eye, in that contact is first established between the sensory organ itself and the brain, which is the interpreter of its signals. A bubble is tied off on either side of the rear portion of the brain. This bubble eventually becomes the inner ear, containing the organs of both hearing and balance. Somewhat later the outer ear, with the auditory canal and the outside of the eardrum, begins to take shape. The intermediate section, the middle ear, with its auditory bones (the hammer, the anvil, and the stirrup), begins to project inward from the throat. It then takes a long time for all the creases and folds of the outer ear to take their final shape. Minor aberrations in the appearance of the outer ear are usually of no consequence for a child's health, hearing, or development.

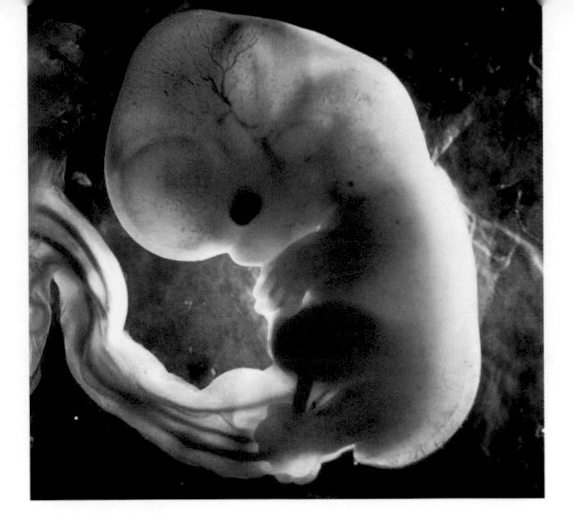

The umbilical cord has three blood vessels: one vein and two arteries.

Circulatory systems meet

Early on the embryo develops a primitive circulatory system, which gradually enlarges to allow blood to circulate out through the aorta to the thin little capillaries in the organs and then back through the liver to the heart. All of the oxygen and nutrients the embryo needs are extracted from the mother's blood. The exchange takes place in the placenta.

The placenta grows to keep pace with the embryo and provides these vital supplies throughout pregnancy. The placental cells develop entirely out of the embryo and share its unique genetic code. The placenta serves as the lungs, intestines, liver, and kidneys for the fetus for nearly the entire time it is in the womb, as the fetal organs do not develop fully until the very end of pregnancy.

Between the placenta and the embryo runs the umbilical cord, which contains three vessels: one big thick one that carries the oxygenated, nutrient-rich blood to the embryo's heart, and two smaller ones that carry the oxygen-poor waste-product-containing blood to the placenta. The umbilical cord coils like a spring because the three vessels are longer than the tube that encloses them—an ingenious safety feature for a lively fetus.

The embryo's red blood corpuscles form in the yolk sac, as do stem cells that will later give rise to white blood corpuscles, the foundation of the fetal immune system. By week 11, the yolk sac is depleted and corpuscle production moves to the liver and spleen. Before long, corpuscle production also begins in the fetal bone marrow. The white corpuscles from the bone marrow then travel to the lymph nodes and thymus to mature.

Whatever food the mother consumes she shares with the embryo through her circulatory system. Blood pulsates plentifully through the umbilical cord, transporting everything the embryo needs. Before reentering the fetal bloodstream, blood from the placenta is filtered through the liver, visible here as a large red shadow below the hand.

(overleaf) In the tiny capillaries of the placenta, the blood corpuscles crowd tightly, waiting to be charged with oxygen and calorie-rich substances from the mother's blood. Once the corpuscles are recharged, they transport their load out to the organs of the embryo. During the fetal stage the corpuscles have a core, which vanishes after birth.

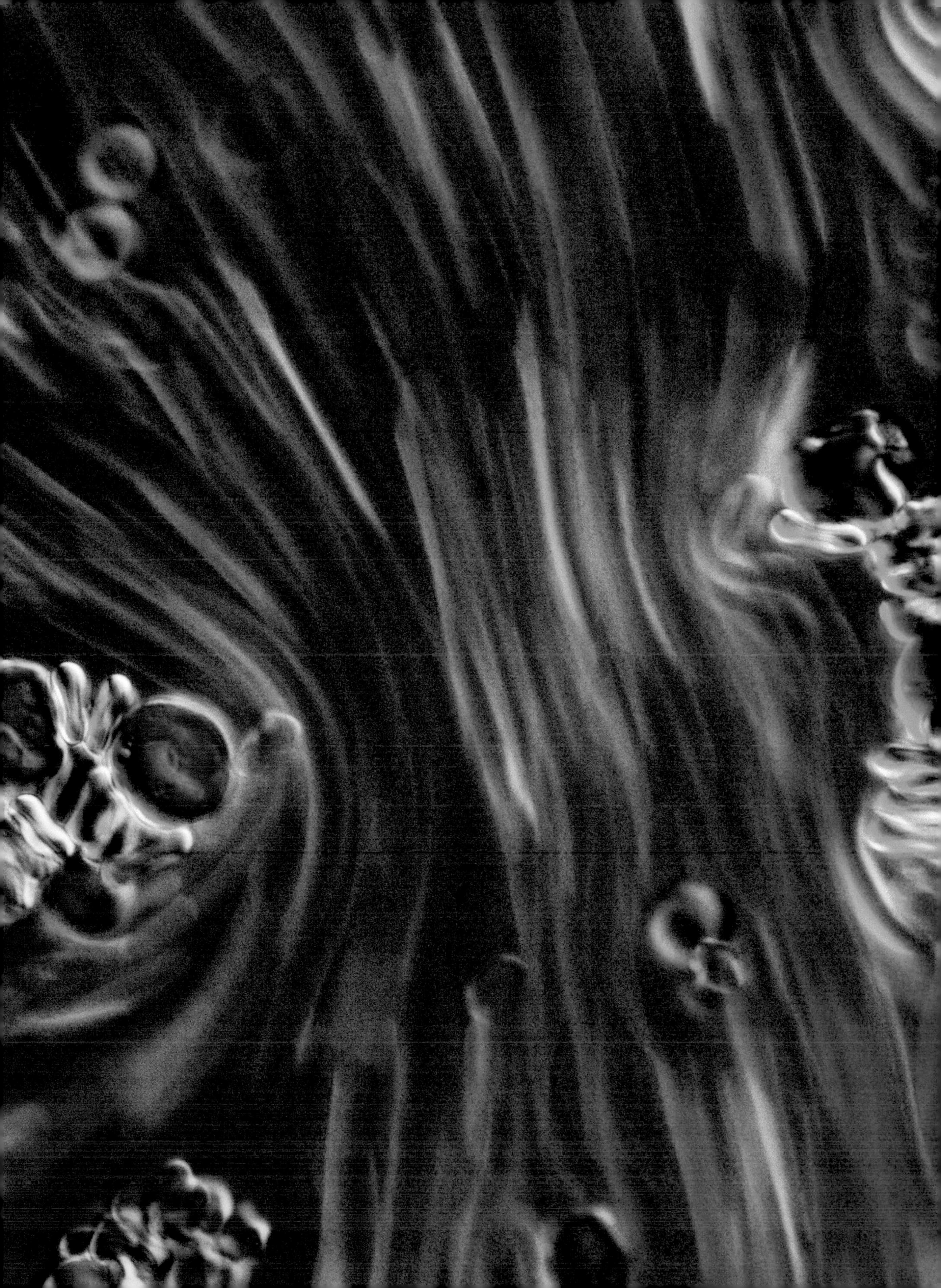

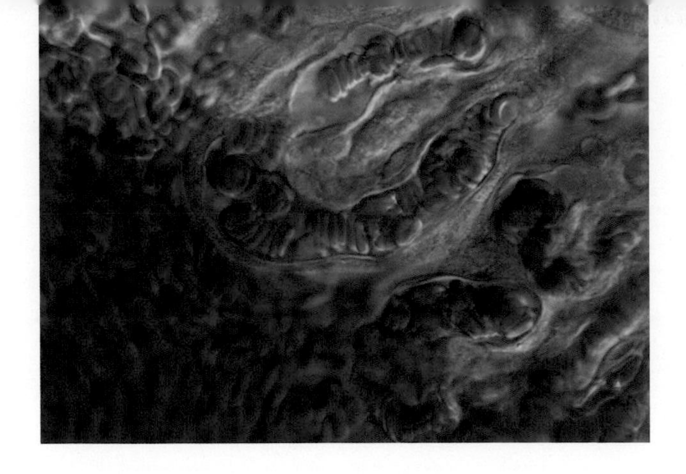

Red blood corpuscles in the capillary vessels of the placenta.

Crucially, the circulatory systems of the fetus and mother remain separate, although a small number of fetal cells do cross into the mother's circulation. Doctors use these cells and their DNA to gain genetic information about the fetus. Some substances that the mother consumes, including some medications, do not pass through the filter of the placenta and affect the mother only. Other substances pass readily through the placental filter and can hurt the embryo or fetus. Alcohol is, of course, the classic example.

Normally the fetus is somewhat protected from bacteria and viruses by the placenta, which provides a barrier to infection. This protection is particularly important during the first half of pregnancy, when the fetus does not yet have a well-developed immune system and when its developing organs are more easily damaged. Although the fetus's immune system has begun to work by the fifth month of pregnancy, it will take a long time to be fully mature.

The placenta also has another job: to produce hormones, including progesterone, estrogen, and hCG. The latter is present in the body only during pregnancy. It occurs at high levels for 8 to 10 weeks and gradually decreases thereafter. Estrogen increases blood flow to the womb, helping it grow. Estrogen production increases by several hundred percent during pregnancy. As early as seven to eight weeks after the woman's last menstrual period, the placenta produces all the necessary hormones; the ovaries are no longer needed for that purpose. Hormone production by the placenta remains essential for the normal continuation of pregnancy and for the fetus to develop "according to plan."

Red blood corpuscles are highly malleable. They can squeeze through the smallest capillaries to deliver oxygen to every inch of the body.

> In the placenta, the circulatory systems of mother and embryo meet, but their blood does not mingle. Blood from the embryo passes through the placenta and returns charged with fresh oxygen and enough proteins, fats, and sugars to keep building cells at a frenetic pace.

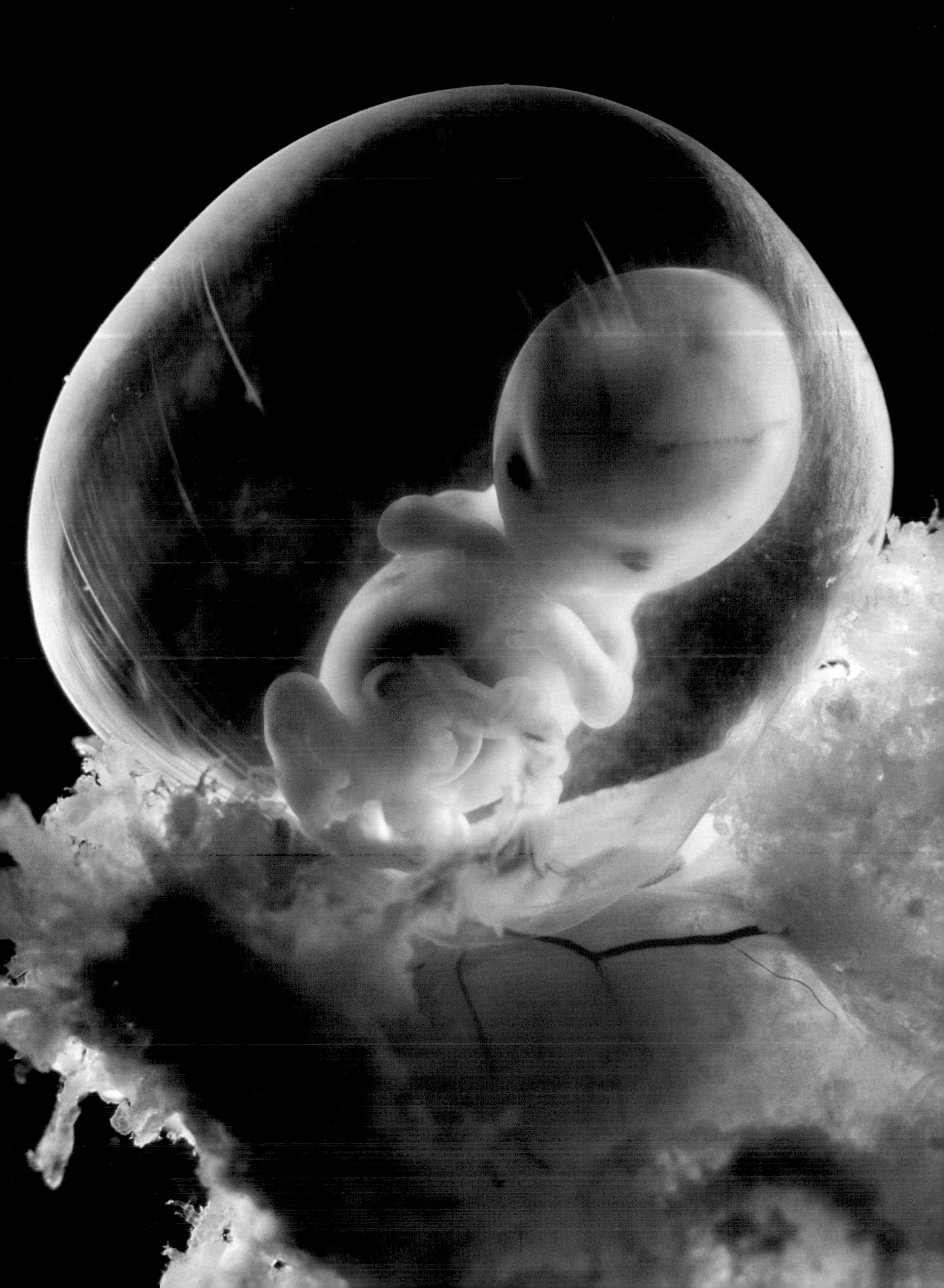

The embryo becomes a fetus

As the expectant mother enters the tenth week of pregnancy, the embryo reaches an important stage of development. The blueprint is complete. All of the organs have been established. The heart has been beating for a month, and the muscles of the torso, arms, and legs have begun to exercise. The embryo, now called a fetus, is about 30 millimeters (1.2 inches) long and weighs between 10 and 15 grams (0.3–0.5 ounces), excluding the placenta and amniotic fluid. In 56 days, it has developed from a single cell into many millions of cells, all precisely programmed for their specific assignments.

If all has gone well, as it usually has, the risk that any serious defect or deformity will develop is now very small. The risk of a miscarriage is also substantially reduced. The pregnancy enters a more stable phase, and for the fetus, now with very pronounced human features, the remaining task is to grow and refine its abilities in preparation for the world outside the uterus. It still has several months left in which to continue developing in the protected environment of the womb.

By this point in the pregnancy the risk of miscarriage is much reduced. Many parents choose this time to share news of the pregnancy with friends and loved ones.

> Week 12. The embryonic stage is over and the fetal stage has begun. The job of the fetus is to build on the embryonic blueprint, refining functionality and doing vital systems testing.

Comfortable or miserable?

During the first few months of pregnancy, most women's breasts are tender, and many women suffer from fatigue. The breasts may begin to change as early as a few days after ovulation. Many women have very little appetite or feel nauseous, especially in the morning. These unpleasant sensations may begin after just a few weeks of pregnancy. The woman may feel she cannot possibly eat the big breakfast she usually enjoys, or that her coffee tastes peculiar. She may be highly sensitive to all kinds of smells, particularly cooking odors. Many women become intolerant of the smell of cigarette smoke.

Some women are so nauseous for the first two to three months that they lose rather than gain weight, both because they have difficulty getting anything down and because they tend to vomit up whatever they do manage to eat. Very occasionally a pregnant woman may need to be hospitalized for a few days to break a vicious circle. There is no really satisfactory explanation for the phenomenon of severe morning sickness. In fact, a woman may be extremely nauseous in one pregnancy and hardly at all in the next. The hormonal changes in her body are surely a significant factor, but anxiety and uncertainty in her new life situation are also aspects to be considered.

What should an expectant mother eat, and how much? One general guideline, for the entire course of a pregnancy, is to eat frequently and not too much at a time. Eating often keeps the blood sugar at an even level, which is an important way to avoid faintness. Eating a little at a time prevents stretching out the stomach, thereby avoiding putting pressure on nearby organs, which may be one cause of unpleasant sensations and nausea.

Many women feel tired and nauseous during early pregnancy. Tender breasts, sensitivity to smells, and depression are also common during this time. Hormonal changes can affect both body and spirit.

Embryonic development follows a detailed pattern day by day. This pattern is determined by the genetic code, yet even before birth there is an intimate interplay between nature and nurture. An unfavorable uterine environment may mean an increased risk of illness later in life. For instance, inexplicably low birth weights in boy babies carried to term have been related to reduced sperm production and impaired fertility in adult men.

Although the placenta plays an important role in protecting the developing embryo and later the fetus, it is not infallible. Drinking, smoking, and certain medicines are known to have a negative impact on the uterine environment and thus on the baby-to-be; so too may maternal deficiencies of nutrients, vitamins, and minerals. The fetus that receives overdoses of sugar and fat may be overnourished, potentially causing damage that, if not visible at birth, may appear, more frequently than was previously thought, later in life.

The issue of what foods are "dangerous" or unsuitable for a pregnant woman is much discussed. Although this kind of advice may often be useful, it is important not to be fanatical. The vast majority of women can go on eating approximately as they normally do, although it is always wise to heed general warnings about foods or additives that could be a health hazard either to the mother or to the fetus. Be cautious about eating raw, smoked, or marinated meat or fish that might contain bacteria. Unpasteurized milk products, such as mold-ripened cheeses, also belong in the less-appropriate-foods category. Avoid large predator ocean fish, such as swordfish, shark, or king mackerel (which may have high concentrations of mercury) as well as freshwater fish (which may have high levels of PCBs or dioxins).

Vegetarians and vegans may continue eating as usual but should be sure to get enough calcium and iron. Medical staff can provide advice about diet and any necessary supplementary vitamins and minerals.

Factors that can affect pregnancy

Drinking, smoking, and drugs

The safety of consuming alcoholic beverages in pregnancy has been much debated, and today the consensus is that alcohol consumption carries a risk of fetal damage.

Children of women who have been heavy drinkers during pregnancy often have a particular appearance, such as severe squinting. But most fetal alcohol damage, known as fetal alcohol syndrome, is invisible, including irreversible brain damage. An occasional drink in early pregnancy, however, is not a reason to terminate a pregnancy.

Narcotic drugs, particularly heroin, are also known to have negative effects and to injure a fetus.

In the United States, about 15 percent of all women who become pregnant are smokers. Despite strong recommendations to the contrary, in the United States a little over 7 percent of women continue to smoke during pregnancy. Women smokers run the risk of having growth-impaired children, and low birth weight is generally considered a risk factor for many illnesses later in life. A link between smoking and childhood asthma has been found, and recent research shows that this risk may arise even in the uterus. This means women should avoid smoking from early pregnancy onward.

Many pregnant women develop an aversion to alcohol and tobacco smoke. Medical practitioners recommend that pregnant women abstain completely from alcohol and tobacco, and this is probably good advice for almost everyone. In Sweden, women are asked about their drinking habits at their first prenatal appointment, and professional counseling is available if needed.

Medications and vaccinations

Some drugs are kept from reaching the fetus by the protective barrier of the placenta; others pass through but are harmless. But a few medicines may both pass through the placenta filter and be harmful to the fetus. The most vulnerable time is during the first eight weeks of pregnancy, when all the organs are being established.

Live vaccines are unsuitable for pregnant women and should be avoided during pregnancy. Flu vaccines, however, are recommended.

X-rays

If a woman of fertile age needs to have an X-ray of the stomach, intestines, gallbladder, kidneys, or lumbar vertebrae, the doctor will always ask whether she is pregnant, as such X-rays are known to be potentially harmful to the embryo. X-rays of arms and legs, mammograms, and dental X-rays, which do not expose the uterus to radiation, may be performed at any time during pregnancy.

At airport security checkpoints, it is normally possible to request a pat-down search instead of an X-ray scan.

Illness and disease

German measles (rubella) is the best known infection that can harm an embryo. If a woman contracts rubella in early pregnancy, the baby's hearing may be impaired, among other things. The onset of symptoms like diarrhea, a high fever, and an itchy rash all over the body could indicate a disease with negative effects on the fetus. With any such illness it is important to contact the doctor for a reliable assessment.

Hazardous substances

There is a great deal of concern today regarding how hazardous substances used in industry and agriculture can harm the uterine environment during pregnancy. We know that some such substances can cause fetal abnormalities, but except in the case of a few well-known toxins, such as PCBs and dioxins, specific links have been difficult to prove. This may be due to the fact that it takes a long time to collect such information, and that it is necessary to study large groups of children to obtain reliable statistics.

Aside from avoiding certain foods, such as certain kinds of fish (see above), it is difficult for individuals to protect themselves from all substances that might pose a risk. However, avoiding exposure to obviously toxic substances, including pesticides, at work and at home is a sensible precaution.

Avoiding exposure to industrial solvents is very important, especially in the first trimester.

Stress and weight

Stress in the woman's external environment is receiving increasing attention as a risk factor in pregnancy. Today stress is considered to influence both miscarriages and early labor.

The percentage of women who are overweight or obese during early pregnancy has risen since the mid-1990s. Obese women have a higher risk of developing complications during pregnancy and childbirth than women of normal weight. This risk is considered at least as great as the risk associated with smoking.

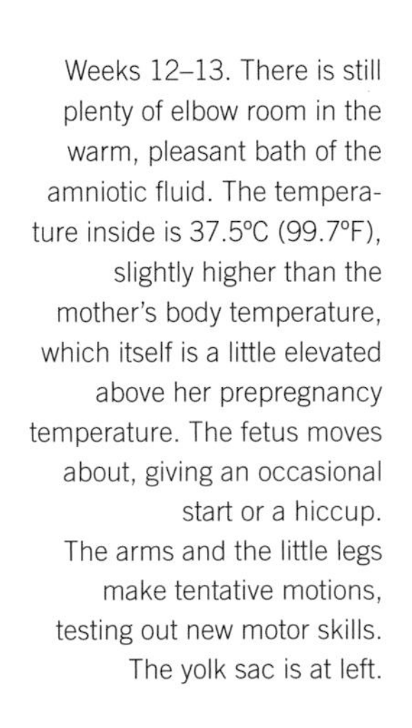

Weeks 12–13. There is still plenty of elbow room in the warm, pleasant bath of the amniotic fluid. The temperature inside is 37.5°C (99.7°F), slightly higher than the mother's body temperature, which itself is a little elevated above her prepregnancy temperature. The fetus moves about, giving an occasional start or a hiccup. The arms and the little legs make tentative motions, testing out new motor skills. The yolk sac is at left.

Prenatal care

Sometime before week 12, the couple will have their first prenatal appointment with a nurse, doctor, or midwife. The parents-to-be are usually very eager and curious, and the medical staff will also have questions for them: How have things been so far? How does the woman feel about her work situation? The answers often provide indirect information regarding how tired the woman has been and how much morning sickness she has experienced. Illnesses, prescription medications, and vaccinations are also important things for the medical staff to find out about for the record. On this or a subsequent visit, they will also discuss lifestyle factors such as BMI (Body Mass Index), a measure of body fat based on weight and height, exercise and diet. Smoking, drinking, and drugs will also be discussed. Today it is considered optimal for both the woman and her partner to receive medical information and to attend prenatal appointments together.

The largest number of questions from both sides will arise in a first pregnancy. When pregnancy is a new experience, the parents-to-be may have lots of questions and some concern about the delivery. Today much time is also devoted to discussing prenatal testing and screening, even though almost all children are born healthy. If the mother is over thirty-five years of age—which is becoming more and more common in developed countries—she will undergo a more extensive examination. After that age there is a slightly elevated risk of chromosomal aberrations in the fetus, particularly Down syndrome. Today many prenatal diagnostic tests can reveal other genetic disorders.

Most parents-to-be are excited about their prenatal appointments. At last they can get answers to all their questions. Doctors, nurses, and midwives are accustomed to sharing in both concerns and excitement, and the appointments usually have a calming effect.

The medical staff will also ask questions about any previous pregnancies. Has the woman had a miscarriage or abortion for any reason? If she has children already, how did her previous pregnancies and deliveries go?

Were there any difficulties? If the woman has had a previous cesarean

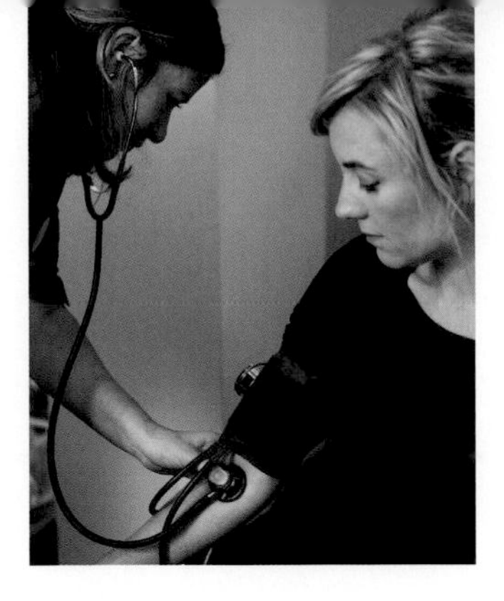

Prenatal visits always include a blood pressure check.

Blood tests provide important information.

section, the staff will need to have as many details as possible. If the cesarean was performed because the pelvis was considered too narrow in relation to the size of the baby, perhaps the baby will have to be delivered by cesarean this time, too; but it is often possible for the mother to have a vaginal delivery even if the last baby was delivered by cesarean.

At the first appointment a urine sample and blood tests will be taken. The blood tests will reveal the woman's blood group and whether she is immune to German measles, has had syphilis, or is positive for HIV or hepatitis. Often genetic carrier testing will be performed for inherited diseases such as cystic fibrosis.

Because the ability of the blood to transport oxygen is so important during pregnancy, it too is investigated, by measuring the hemoglobin levels, or blood count. The blood count is followed regularly for the rest of the pregnancy. Hemoglobin production is contingent on access to iron, so it is very important that a pregnant woman gets enough iron, either through her diet or by taking supplementary iron pills throughout the pregnancy.

Blood pressure is monitored regularly during pregnancy. Women who have high blood pressure before becoming pregnant will need to continue treatment. High blood pressure that develops in the third trimester along with protein in the urine and generalized swelling (edema) is sometimes treated with medication and rest. This condition, called "preeclampsia," may require an early delivery.

Sometimes an ultrasound scan is performed at this time. If there is more than one fetus in the uterus, they will be seen. Twins may share the placenta or have separate ones.

The health of fetus and of mother are closely interrelated. For this reason the first prenatal appointment includes a number of tests. It is important to take the pregnant woman's blood pressure and blood count and be sure there is no protein or sugar in her urine.

> When is the baby due? The doctor or midwife can make an estimate using a pregnancy wheel (or these days an app), but the first ultrasound scan early in pregnancy will establish a more precise date.

NOVEMBER
DECEMBER
JANUARI
FEBRUARI

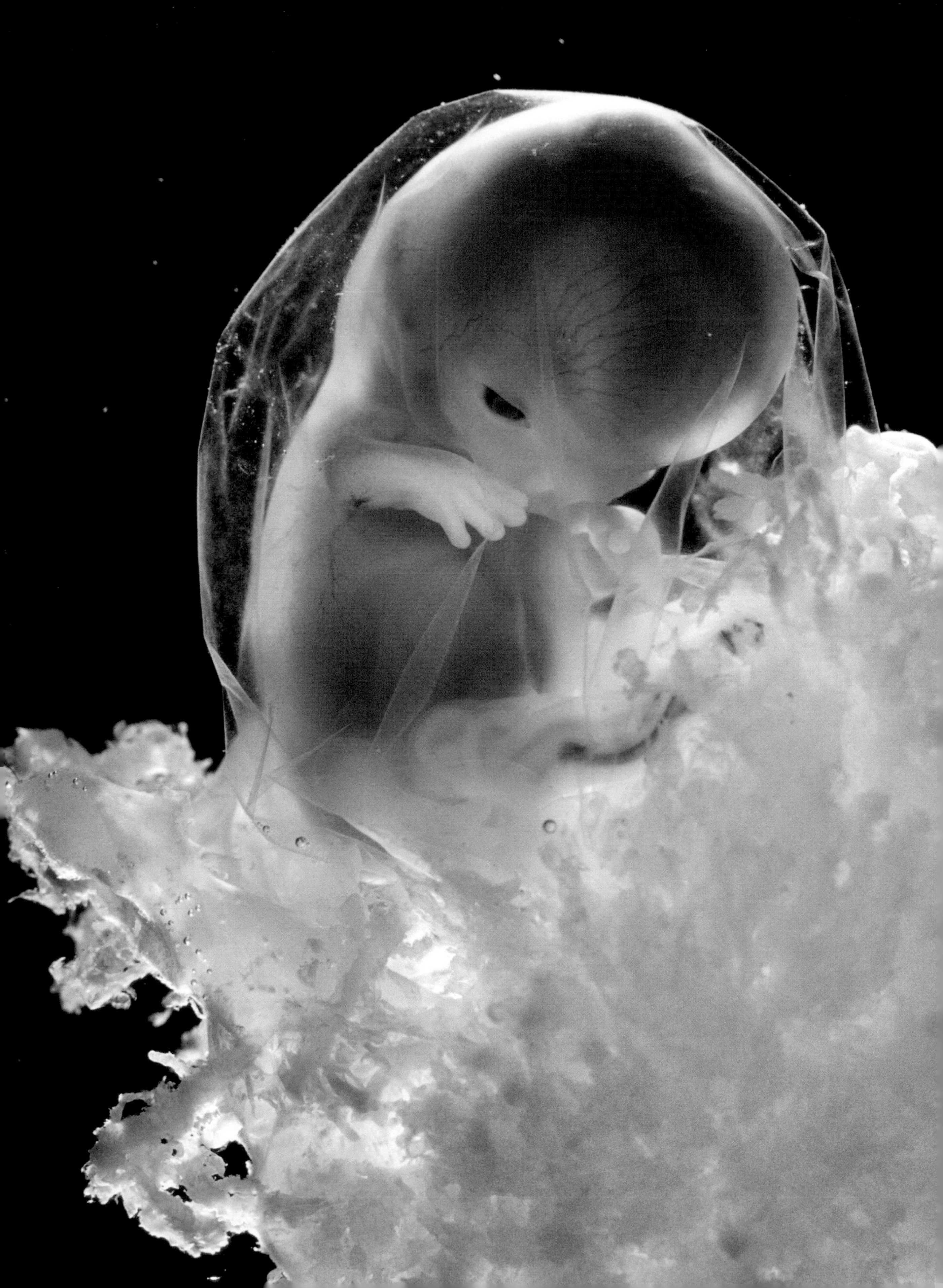

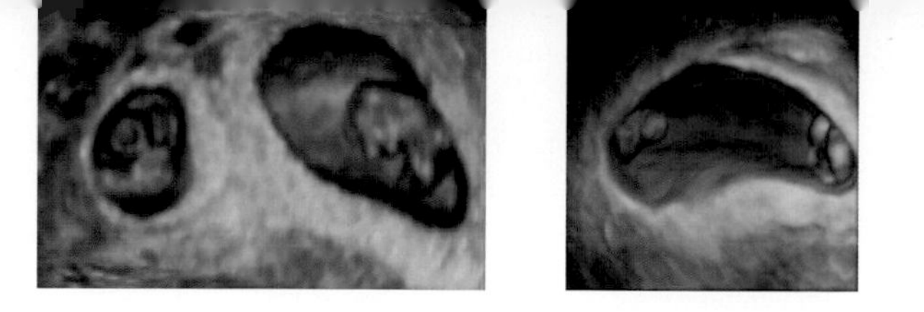

Twins in week 7, left in their own amniotic sac, and right in the same.

Twins may be either identical (monozygotic) or nonidentical (dizygotic). When an egg is fertilized by a sperm, all the cells have the same genetic code. If, after just over a week, the egg divides into two halves, two viable embryos will develop into two individuals with almost precisely the same genetic code. How similar they become will depend on the environment in the uterus. In some families genetic factors govern a predisposition for monozygotic twins.

For nonidentical twins the ovary releases two eggs rather than one, a somewhat more common event as a woman ages. Of course, dizygotic twins may be of the same sex, but they are no more similar than any other siblings. After birth, genetic tests can determine with certainty whether a set of twins is monozygotic, should it be important to make this determination. An ultrasound scan during the first three months can usually also confirm whether twins are mono- or dizygotic; later in the pregnancy this becomes more difficult to see.

It is still difficult to get detailed ultrasound images of individual organs. The presence of a heartbeat is a good sign of life. So is an occasional energetic movement by the fetus or fetuses.

This early in pregnancy, ultrasound cannot show whether the baby-to-be is a boy or a girl. As described earlier, however, the genetic code that determines the sex of the baby was established at the moment of fertilization: two X chromosomes for a girl, one X and one Y for a boy. A particular gene in the Y chromosome, known as SRY, is known to be essential for the development of the fetus into a normal male child. Another important contributing factor in sex differentiation is the cells from the yolk sac, which help develop both internal and external genitals.

The genitals begin to form during the embryonic stage, in weeks 8 and 9, although at that point the sex glands and organs are identical for boys and girls. A little bud develops between the legs; in boys it develops into the penis, in girls into the clitoris. Gradually two little bulwarks take shape on either side of a crack. In a boy, they merge to become the scrotum, while in girls they structure the vaginal walls, and the crack will not fuse shut.

If there is any uncertainty about how long the woman has been pregnant, the first appointment is often concluded with an ultrasound scan. This scan will also show if the woman is carrying twins. Most twins have individual amniotic sacs, but in rare cases they may share the same sac as well.

◁ Fetus, week 10.

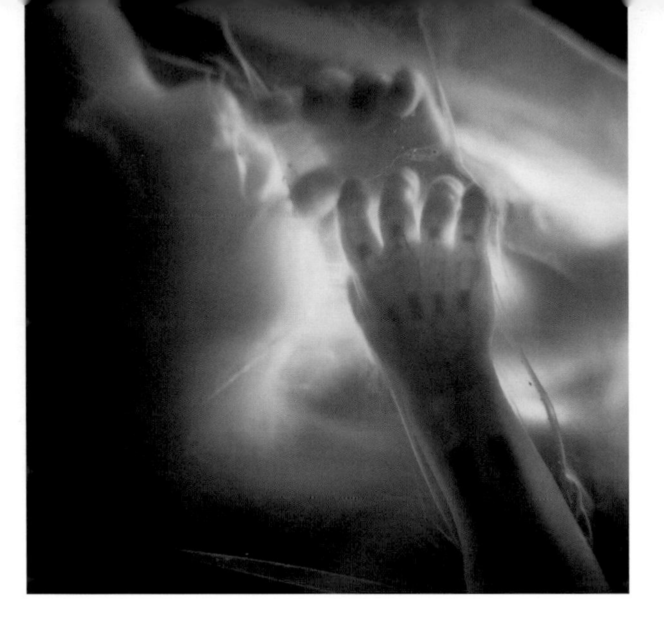

The cartilage has become calcified in the ossification centers outlined in the bones of hand and arm.

Growth spurt

When the expectant mother enters the fourth month of pregnancy, the most vulnerable phase of the pregnancy is over: after the twelfth week miscarriage is relatively infrequent. In the uterus, surrounded by amniotic fluid, with a heart that beats twice as fast as that of its mother, a human-being-to-be who has just been through the first prenatal exam is napping. The fetus has settled in. It is still growing fast, and the uterus expands to accommodate it. By the end of the twelfth week it is the size of a man's fist, and just a month later it has grown to the size of a honeydew melon. The placenta too is becoming bigger and thicker, in step with the fetus's demands for oxygen and nutrition.

All the organs that were created during the embryonic stage are developing and growing, and the proportions of the little fetus are becoming more and more like those of a newborn. The head is still disproportionately large, about one third of the total body length. The face continues to develop. By 11 or 12 weeks of age, its features look distinctly human. The gelatinous body is becoming firmer, as cartilage is transformed into bone. Cartilage is softer than bone and more malleable.

The skeleton is first made of cartilage, then is gradually transformed into bone according to very specific patterns, as long tubes become arms and legs. Calcification begins in the midsection of a bone, while the growth zones close to the ends of the bone remain soft and are not completely ossified until adolescence. Once cartilage becomes bone, lengthwise growth is halted.

Week 15. The fetus grows constantly, so its body must be malleable.Therefore the skeleton is first made of cartilage, then gradually ossifies into bone. The calcification process begins in the long arm and leg bones.

> In the third month, facial features emerge: the forehead grows in size, with the branching blood vessels clearly visible through the translucent skin. The eyelids are closed and will not open again until month seven.

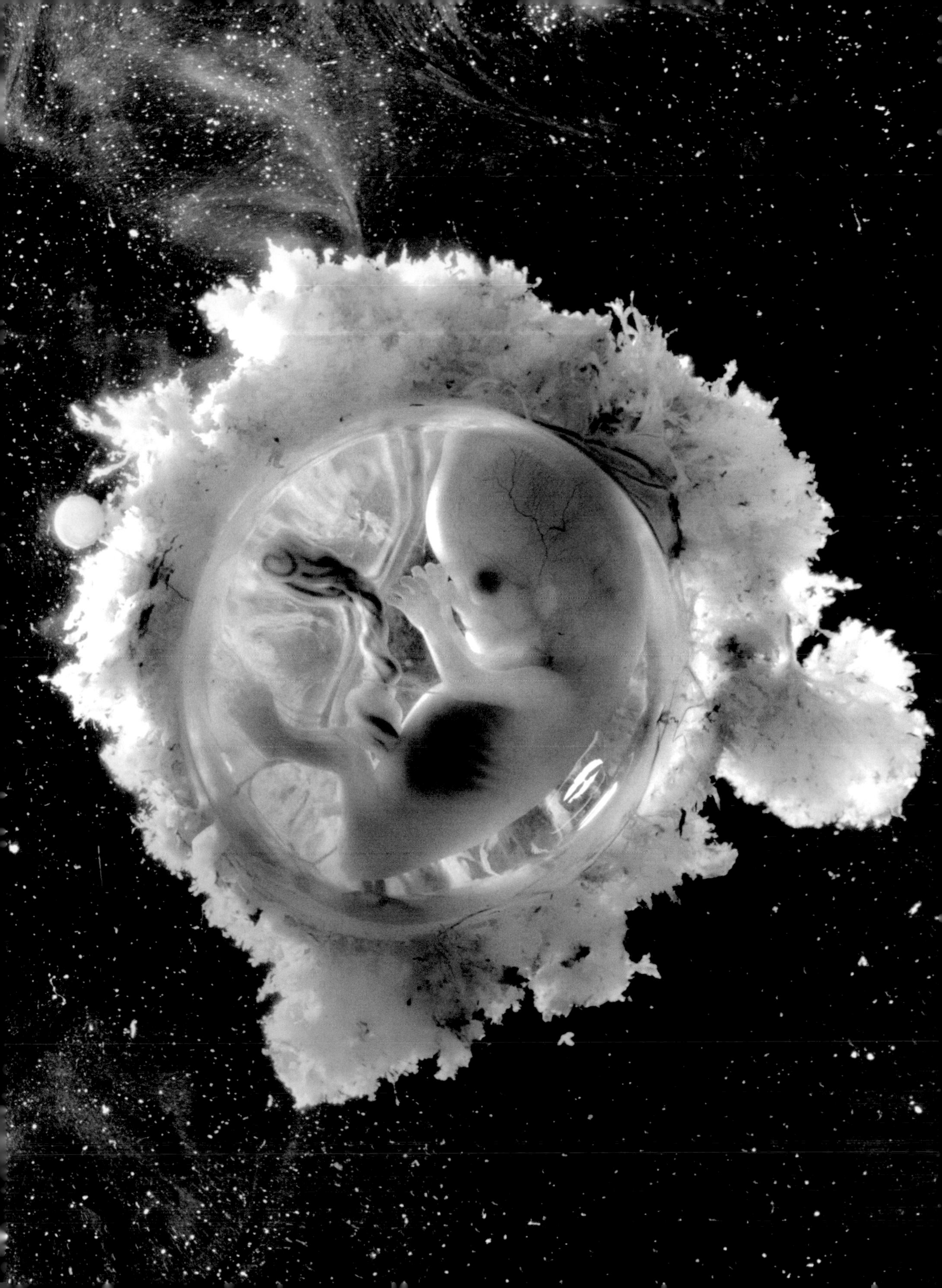

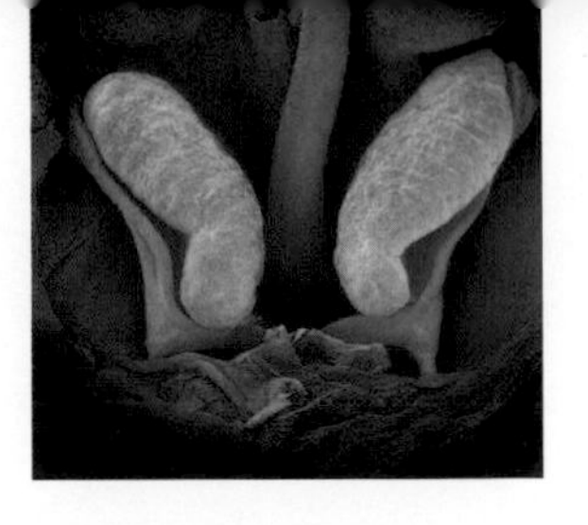

Ovaries in week 11.

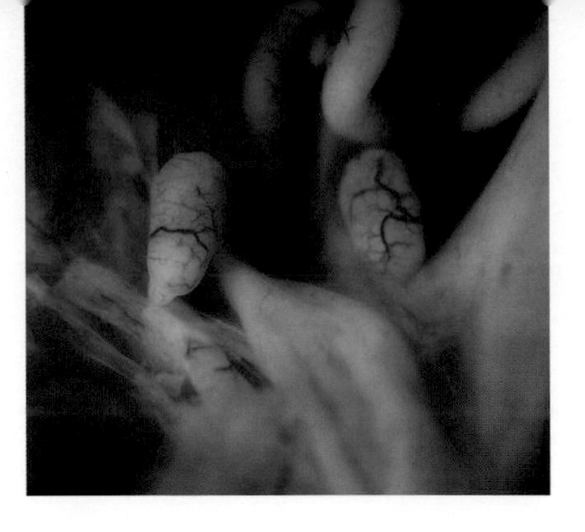

Testicles in week 13.

The extremely sensitive brain needs a protective shell, but it must be allowed to grow. In the fetal stage the skull consists of large, bulging bones that are loosely connected, so the brain has plenty of room to expand.

The little fetus moves more and more every day, and the jerky body motions during the embryonic stage are now replaced by slower and apparently more goal-oriented movements. The hands often find their way to the mouth, ultrasound scanning shows, and the arms and legs stretch and bend. The occasional breathing movement also appears; the fetus can be seen to yawn or hiccup now and then and seldom lies perfectly still for any length of time. The pattern of movement is much the same day or night; not until later in pregnancy will the fetus display a more regular day-and-night rhythm, with extended periods of sleep. Early on it never seems to sleep for more than a few minutes at a time.

The external genitals have begun to grow. Some expectant parents are eager to know before the birth whether their baby will be a girl or a boy, while others prefer to remain in suspense. Today it is possible to determine the sex of a baby during pregnancy, most easily by ultrasound scan. Sometimes, however, it can be hard to see the sex, depending upon the position of the fetus in the womb. Placental sampling or amniocentesis can also reveal the sex of the baby, as can a check of maternal blood for fetal DNA, "cell free DNA."

Even before the genitals are fully formed, the ovaries of the female fetus already hold several million immature eggs, while in the male fetus, the testes are producing testosterone and immature sperm.

> Cartilage in the toes is transformed into bone. Once the toes are under way, the legs have their turn. By week 12 the fetus is in constant motion, but its movements are still too tiny for the mother to feel.

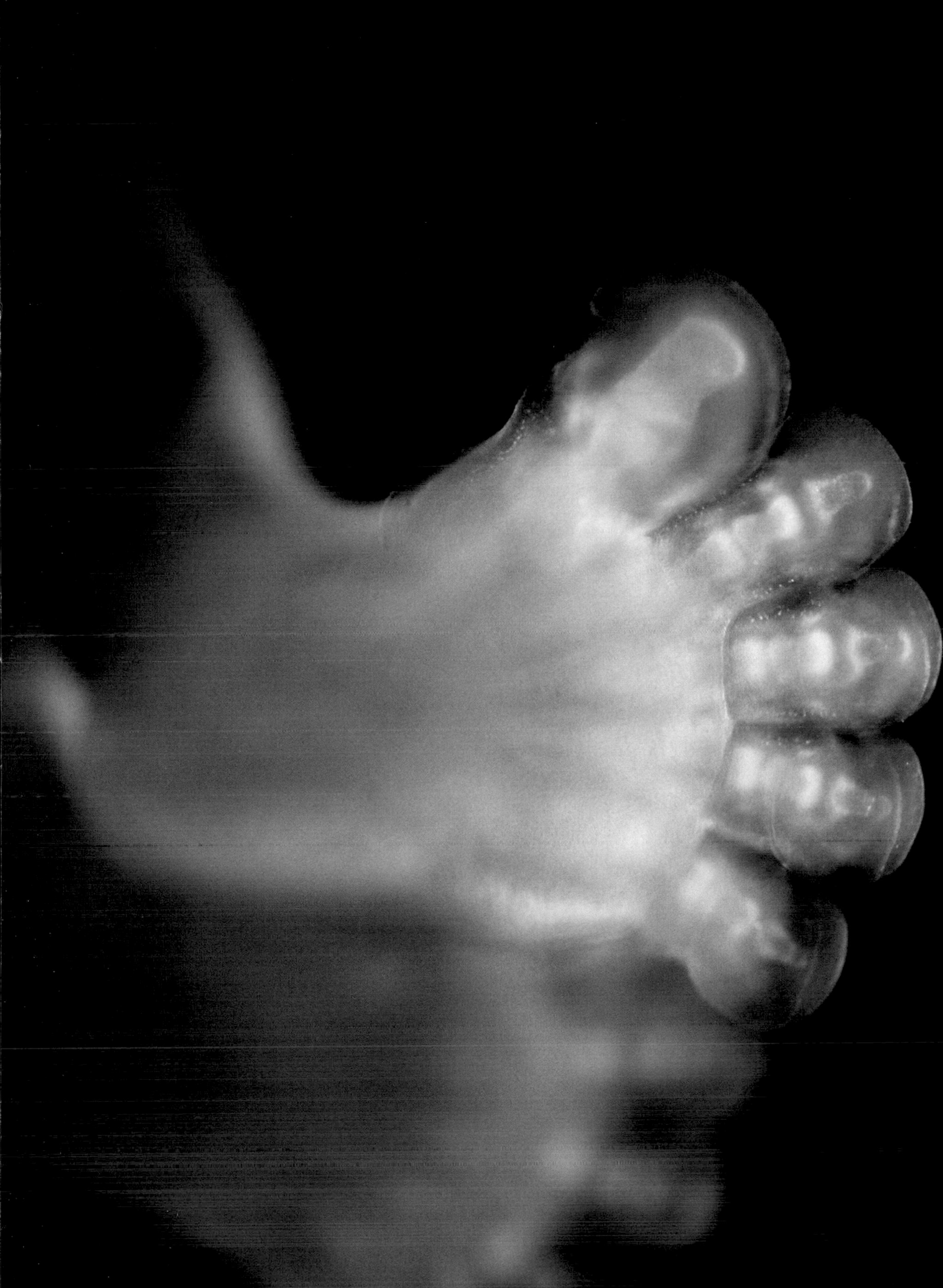

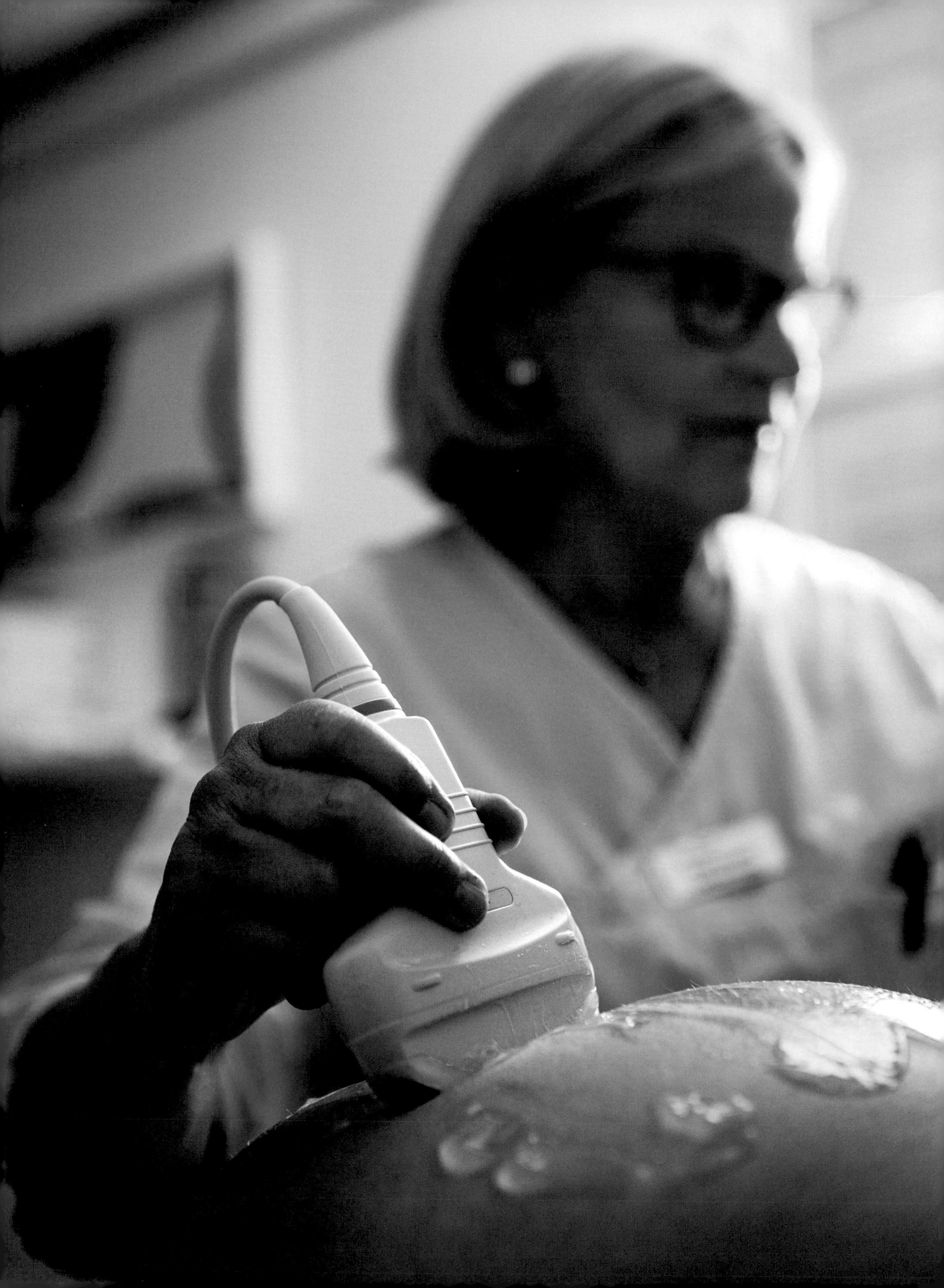

Everything all right in there?

Today expectant parents often express overt concern, more frequently than in previous generations, that things might go wrong during a pregnancy and the baby might not be healthy. This is probably largely due to our increased knowledge about the causes of deformities and other disabilities. An early miscarriage is nature's way of ensuring that an embryo with seriously damaged genetic material will not go on developing. But if the damage is less severe, the pregnancy will continue, and the parents may face a difficult decision.

Three percent of all babies born have some little aberration or deformity, such as a deformed outer ear or a disfiguring birthmark, which may upset the parents and affect the child's appearance. In about one percent of these cases, however, the deformity is more severe, such as a heart defect, or an abnormality of the abdominal or intestinal system or of the urinary tract. In such cases surgery will be necessary. Spina bifida is a deformity that often requires repeated operations and long periods of hospitalization and produces a great strain on both child and parents.

The most common genetic disorder is the one that causes Down syndrome. Children with this syndrome have an extra chromosome in pair 21, and thus have forty-seven instead of forty-six chromosomes. Information from the extra chromosome influences development, resulting in superficial external anomalies (slanted eyes, deformed ears, and abnormal hands) and many internal ones, affecting intellectual ability among other things. The risk of having a child with Down syndrome is greater in older women, but it occurs even with young couples. Because the majority of all babies are born to younger parents, most parents of children with Down syndrome are young enough to be in an age group that, in most countries, does not routinely receive prenatal diagnostic testing.

During the early months of pregnancy ultrasound examinations are usually performed vaginally, but as the fetus grows, an external examination, performed by running the transducer over the mother's abdomen, becomes more informative. The gel helps eliminate air pockets, making the onscreen image clearer. Sometimes both kinds of examination will be performed, depending on what information is needed.

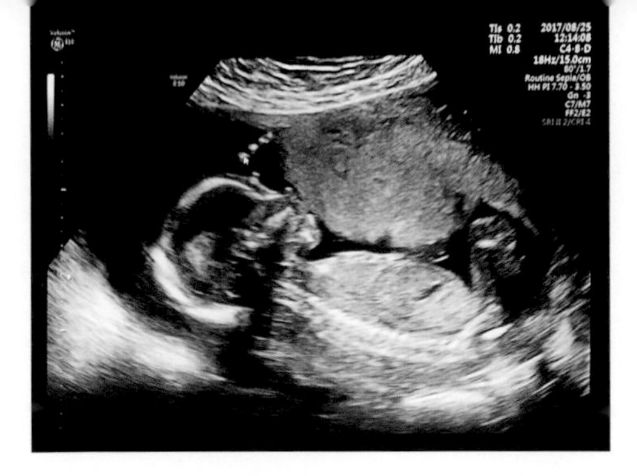

Ultrasound scan in week 18.

In any case, not all couples who receive this diagnosis decide to terminate the pregnancy. Parents and siblings of children with Down syndrome have often testified about how these children enrich the lives of the other family members. Moreover, the degree of mental disability associated with the syndrome varies widely. It is important that the woman and her partner are provided with all the information they need to make a well-founded decision.

Today many types of prenatal diagnostic tests make it possible to determine whether the developing child has a congenital disorder. The most common is the ultrasound exam, of which many women have several during their pregnancies. Others include combined or first trimester screening (a combination of several tests to calculate the statistical level of risk of birth defects), noninvasive prenatal testing (NIPT), chorionic villus sampling, and amniocentesis. But the new opportunities to discover even minor aberrations may pose problems for physicians as well as for expectant parents, who may be faced with a life-changing decision. Although prenatal tests sometimes lead to definite conclusions, at other times it is impossible to know precisely how a genetic aberration found in utero will affect the child when it is born.

Most pregnant women who have prenatal tests, particularly ultrasound, consider the testing to be a way of obtaining extra reassurance that all is well and have not usually worked through how they might react if some disorder was actually found. For this reason, physicians should always discuss the possible consequences of such examinations thoroughly with the expectant parents beforehand. Some kinds of testing may carry a slight risk of miscarriage, and the couple must be informed of that possibility.

From a medical point of view, an ultrasound scan is done to see if there is anything out of the ordinary. The physician studies the fetus carefully, explaining what is seen in the picture: the big, rounded head, the rapidly beating heart.

> Seeing one's unborn child appear on an ultrasound screen, a shadowy form with floundering arms and legs, is a major experience. This may be the moment—with photographic evidence in hand—when the expectant parents fully realize what is in store for them.

Prenatal tests

Ultrasound

Many countries require that every expectant mother be offered at least one ultrasound scan, including organ screening, during her pregnancy. In practice, though, ultrasound has come to be used much more frequently, now that it is known to be relatively harmless.

The benefits of an ultrasound scan clearly outweigh any possible small risks, but it is wise to exercise moderation and perform tests only when they are medically justified.

A state-of-the-art ultrasound monitor can show the head, chest, beating heart, torso, arms, and legs as early as week 11. The fingers can be counted and the genitals seen. In most cases it is possible to see if the baby is going to be a girl or a boy in week 17.

An ultrasound scan in the second trimester, sometime in weeks 18–19, is routine. It allows the physician to calculate gestational age, check the position of the placenta, see if the mother is expecting twins, examine the anatomy of the fetus or fetuses, and check for the presence of some birth defects, including Down syndrome.

Combined screening

Also known as "first trimester screening" in the United States, this is a common and noninvasive method for assessing the risk of Down syndrome as well as abnormalities in chromosome pairs 13 and 18. A blood sample taken from the mother no earlier than week 9 is tested for human chorionic gonadotropin (hCG) and pregnancy-associated plasma protein A (PAPP-A). Then in week 12 or 13, an ultrasound scan is done to measure the size of the fetus and the amount of fluid behind the neck. This is known as a nuchal translucency test. The age of the mother is noted. If chromosomal aberrations were detected in any of her previous pregnancies, this is recorded as well.

All of this information is fed into an algorithm, which returns a statistical risk of abnormality. Depending upon the calculated risk factor, additional tests may be warranted. These may include NIPT, chorionic villus sampling, or an amniocentesis. Naturally, whether to pursue further testing is a decision for the parents to make together.

Noninvasive prenatal testing

Noninvasive prenatal testing (NIPT), or cell free DNA, is another method of calculating the probability of a chromosomal abnormality. It uses a blood sample taken from the mother in or after week 9. Because some fragments of fetal DNA pass through the placenta into the maternal bloodstream, examining the mother's blood can detect aberrations in the fetal chromosomes.

NIPT examines chromosome pairs 13, 18, and 21. It can also reveal the blood type and sex of the fetus. If NIPT results suggest a chromosomal abnormality, this will need to be confirmed by amniocentesis.

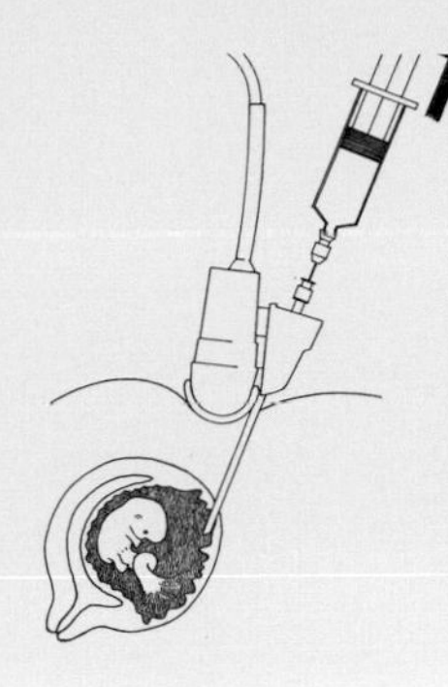

Chorionic villus sampling.

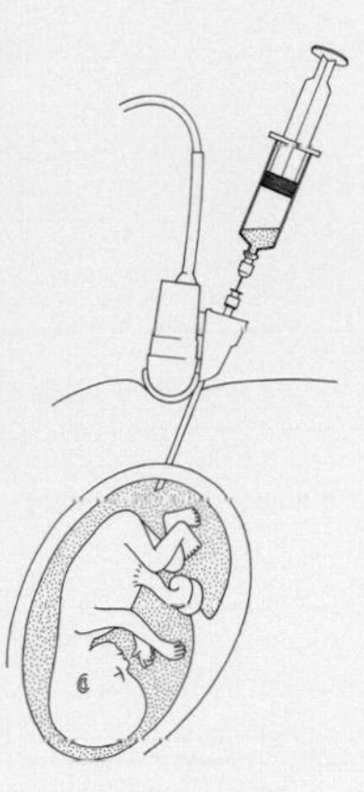

Amniocentesis.

Chorionic villus sampling

Chorionic villus sampling (CVS) passes a thin needle into the placenta, guided by ultrasound, to extract a tiny sample of tissue from the chorionic villi. Villi are tiny fingers that exchange oxygen and waste between mother and embryo. By sampling them—their genetic makeup is identical to that of the embryo—it is possible to determine the genetic code of the baby-to-be. Very occasionally, the chromosomal material of the placenta and embryo may not match exactly, which can make interpreting the results of this test more difficult.

CVS is usually performed during the tenth or eleventh week and there is a very low risk of miscarriage as a result of the test.

Amniocentesis

Amniocentesis was formerly a common method for directly investigating possible fetal genetic disorders. Today, however, it has largely been replaced by combined and NIPT screening, which carry no risk to the fetus. Still, the risk of an amniocentesis leading to miscarriage is low, as it is for CVS.

The test is carried out after week 15. Cells shed by the fetus are found in the amniotic fluid; by extracting 10–20 milliliters (3–6 ounces) of this fluid, using a hollow needle and a fine syringe, the fetal cells can then be cultured in the laboratory for a couple of weeks, after which it is possible to determine their chromosomal makeup. Each pair of chromosomes is then examined.

Amniocentesis can also be used to perform targeted genetic tests if the parents are known to carry the genes for certain heritable diseases. And, of course, amniocentesis can indicate whether the fetus is a boy or a girl.

If abnormalities are suspected in chromosome pairs 13, 18, or 21, a rapid genetic test can provide those specific results within a day or two, although it cannot yet give a more complete picture.

Other tests

A Doppler ultrasound can measure fetal circulation, registering the flow of blood through the fetal umbilical artery and the maternal placental arteries to give a picture of the nutrient and oxygen supply to the fetus. Ultrasound can also be used to estimate fetal weight. This can be done continually throughout the later stages of the pregnancy if the doctor suspects that the fetus is not growing at the normal rate.

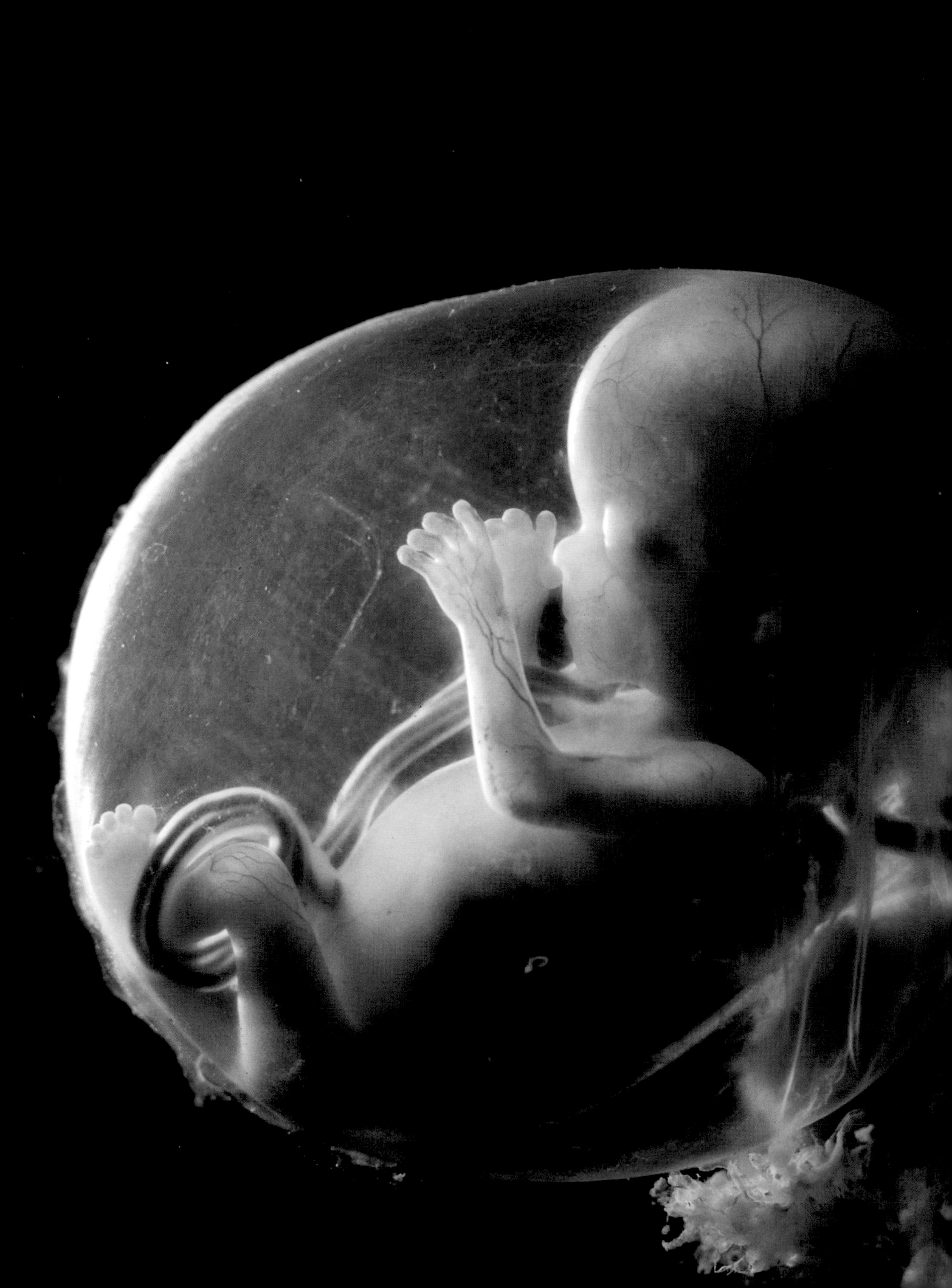

A swirl of hair on the forehead.

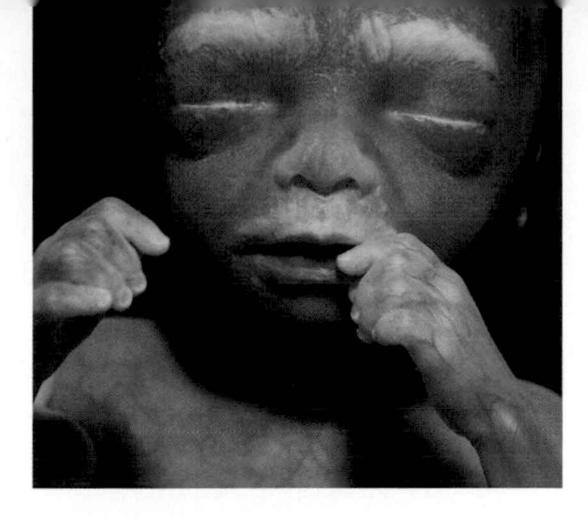

The skin is protected by the waxy vernix.

An aqueous life

The fetus is encapsulated in the liquid-filled amniotic sac. Evidently human beings are constructed to spend the first nine months of our lives in water, just as all animals, at the dawn of evolution, lived in the original sea and became adapted to life on dry land only gradually.

Early in pregnancy the amount of fluid in the sac is negligible, but it increases with every passing day, to allow more freedom of movement for the growing fetus. The liquid is not crystal clear but is full of suspended flakes. The fetus swallows amniotic fluid, which passes through the gastrointestinal system, resulting in waste products. Other flakes are actually little clumps of cells from the lungs—a result of the fetus having swallowed fluid into the pulmonary system and then coughed it up. In addition, superficial cells flake off the skin of the fetus and float in the amniotic fluid. These cells can be used for prenatal diagnostic testing, as described above. The enclosed sea also contains the urine of the fetus, filtered by its kidneys. In spite of all this "pollution," the amniotic fluid is perfectly sterile.

The skin of the fetus is well adapted to an aqueous life, protected by a waxy substance called the vernix. When the skin is more or less fully developed, the very first hairs take shape, as early as the twelfth week of pregnancy. The first hair is known as lanugo, or "woolly hair," and covers the whole body, with a characteristic pattern on the skin. Around weeks 16 to 20, the roots of the hair on the head and in the eyebrows become a little thicker and more distinct. Pigment cells also begin to color individual strands of hair.

The lanugo falls away before the baby is born. Its precise role during the time in utero has not yet been determined, but it may function as a sort of binder for the protective vernix, which is produced in large quantities from the sebaceous glands, located along the roots of the hair. Some of the vernix is rinsed off the skin and mixes with the amniotic fluid, which by the end of the pregnancy has become cloudy. The vernix covers the fetus throughout pregnancy, although it thins toward the end of pregnancy. Babies born after their due date often have little or no vernix left.

The fine lanugo grows in the same pattern as the head hair and helps to bind the vernix, which protects the skin against chafing and minor injuries.

< Week 8. The fetus floats in the amniotic sac, surrounded by amniotic fluid at body temperature.

(overleaf) The lanugo, or "woolly hair," covers the entire head.

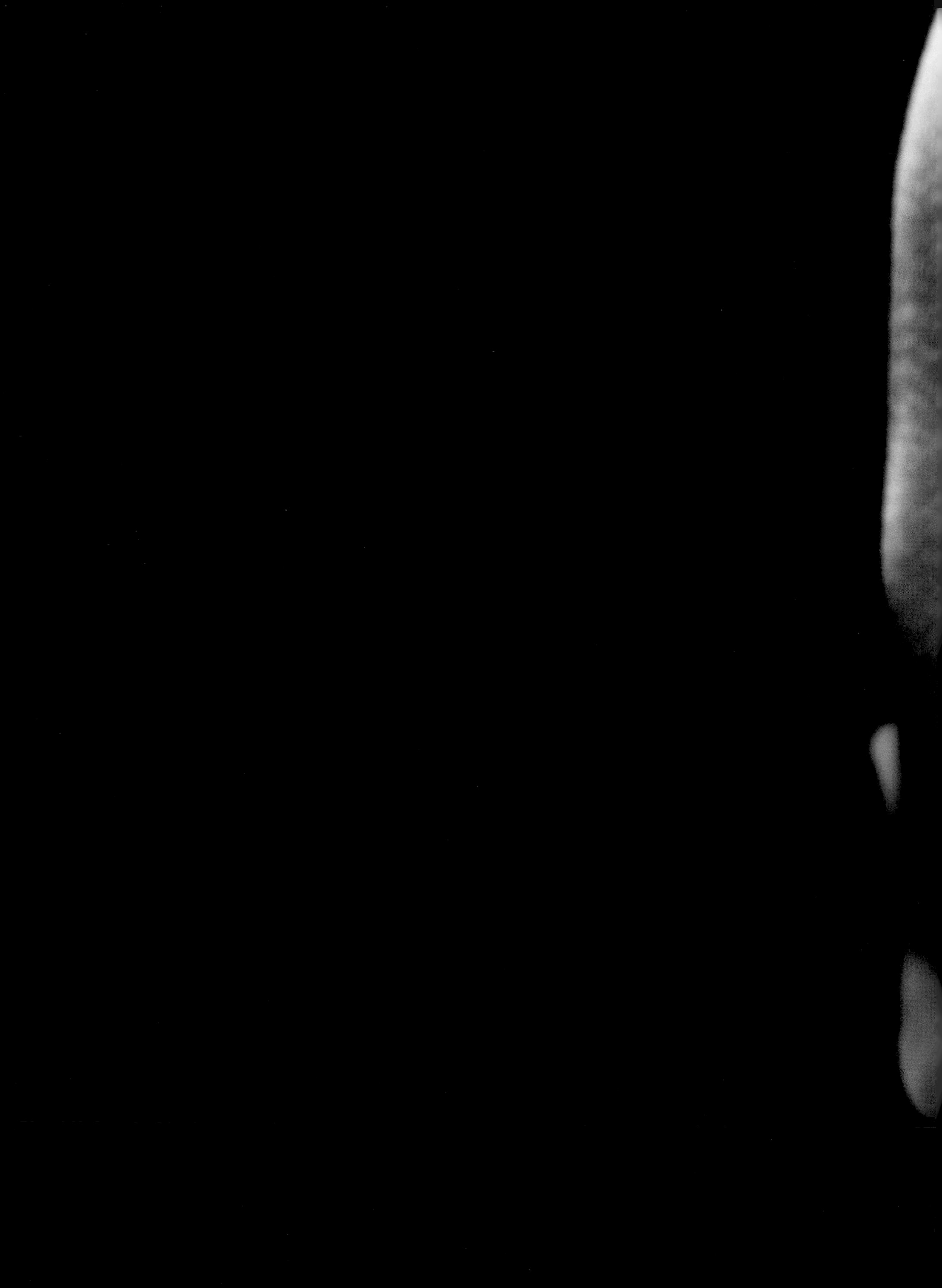

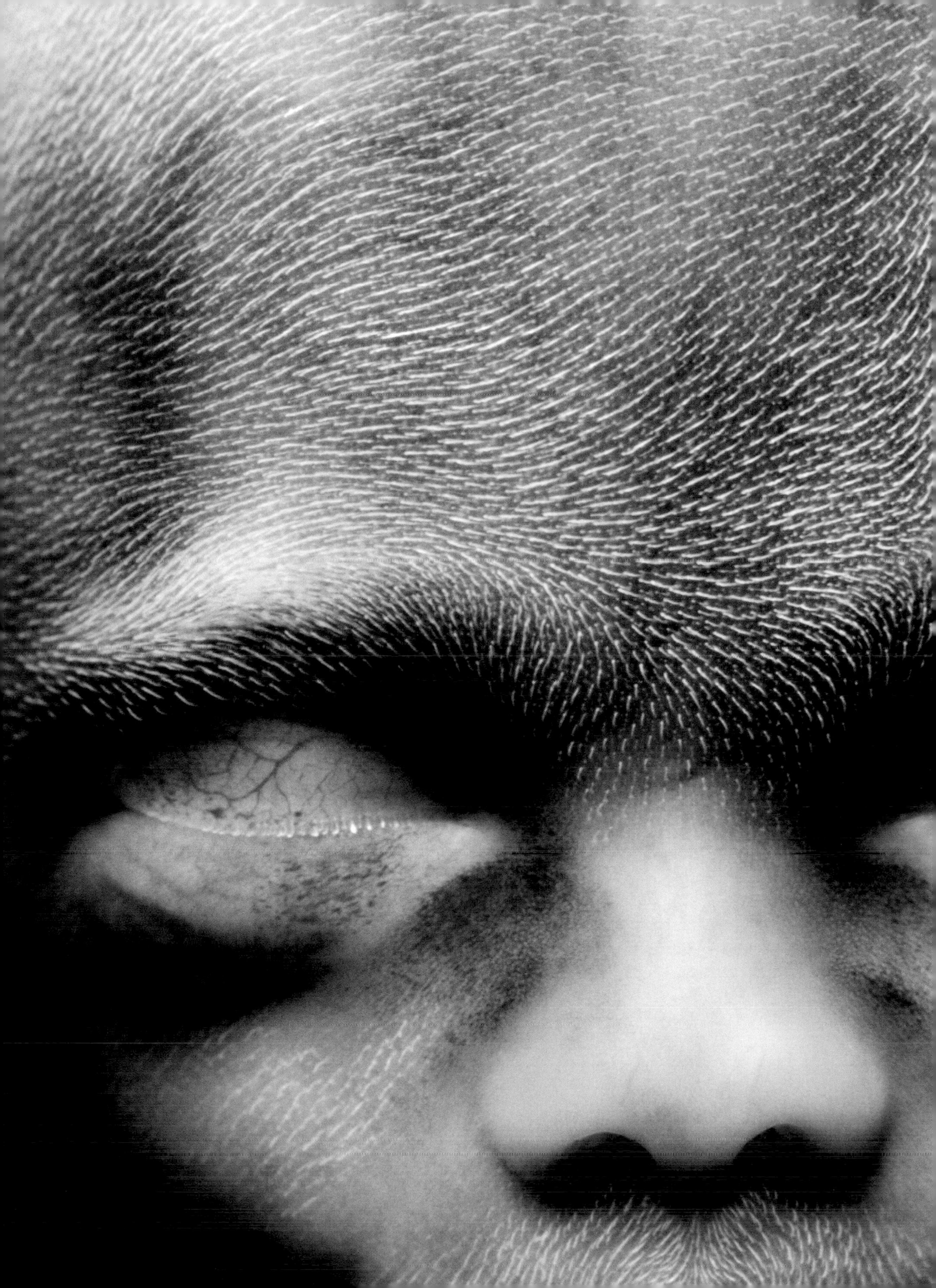

Halfway there

In midpregnancy many women feel particularly well and find themselves in harmonious balance. Women who have had a great deal of nausea often feel better now, and fatigue declines. A pregnant woman's hair, eyes, and skin may appear very shiny and healthy, the size of her abdomen is not yet restricting her movements very much, and her respiratory system is not much affected, as plenty of room is still available in her abdominal cavity. She is sleeping well at night, although she may feel hot.

Although the expectant mother is aware that her abdomen is growing and the ultrasound scan has shown her a little fetus with a heart beating fast, until now she has experienced the fetus as lying completely still. Then one day she senses a soft fluttering in her belly. These first tangible movements may feel like gas bubbles rising and only gradually like definite bumps and kicks. The first signs are nothing a woman can be taught to feel; they have to be experienced to be known. A woman who has been pregnant before will recognize them more easily. As a rule, a woman who has had babies before feels the movements for the first time around week 16 or 17, while a woman expecting her first child is aware of them a couple of weeks later.

During the middle trimester of pregnancy the expectant mother undergoes visible changes, although she likely does not gain very much weight. Many factors contribute to determining how pregnant a woman looks at this point, including her height and her physical condition. In women expecting for the first time, the pregnancy usually begins to show later.

The growth of the abdomen in pregnancy also reflects the development of the fetus, which is why at prenatal visits medical staff regularly measure the distance between the upper edge of the pelvic bone and the upper edge of the rounding of the woman's abdomen (known as the symphysis-fundus measurement). This distance differs from woman to woman and from pregnancy to pregnancy, but the standard measure for week 24 of pregnancy is 20–25 centimeters (7.8–9.8 inches). In a first pregnancy, the distance increases by about 1 centimeter (0.39 inch) per week until the delivery.

Suddenly this invisible family member is a physical presence. The early movements, which feel like fluttering butterfly wings, soon develop into stronger thrusts. Late in pregnancy the baby's kicks may be so hard, they take the mother's breath away or wake her from sleep.

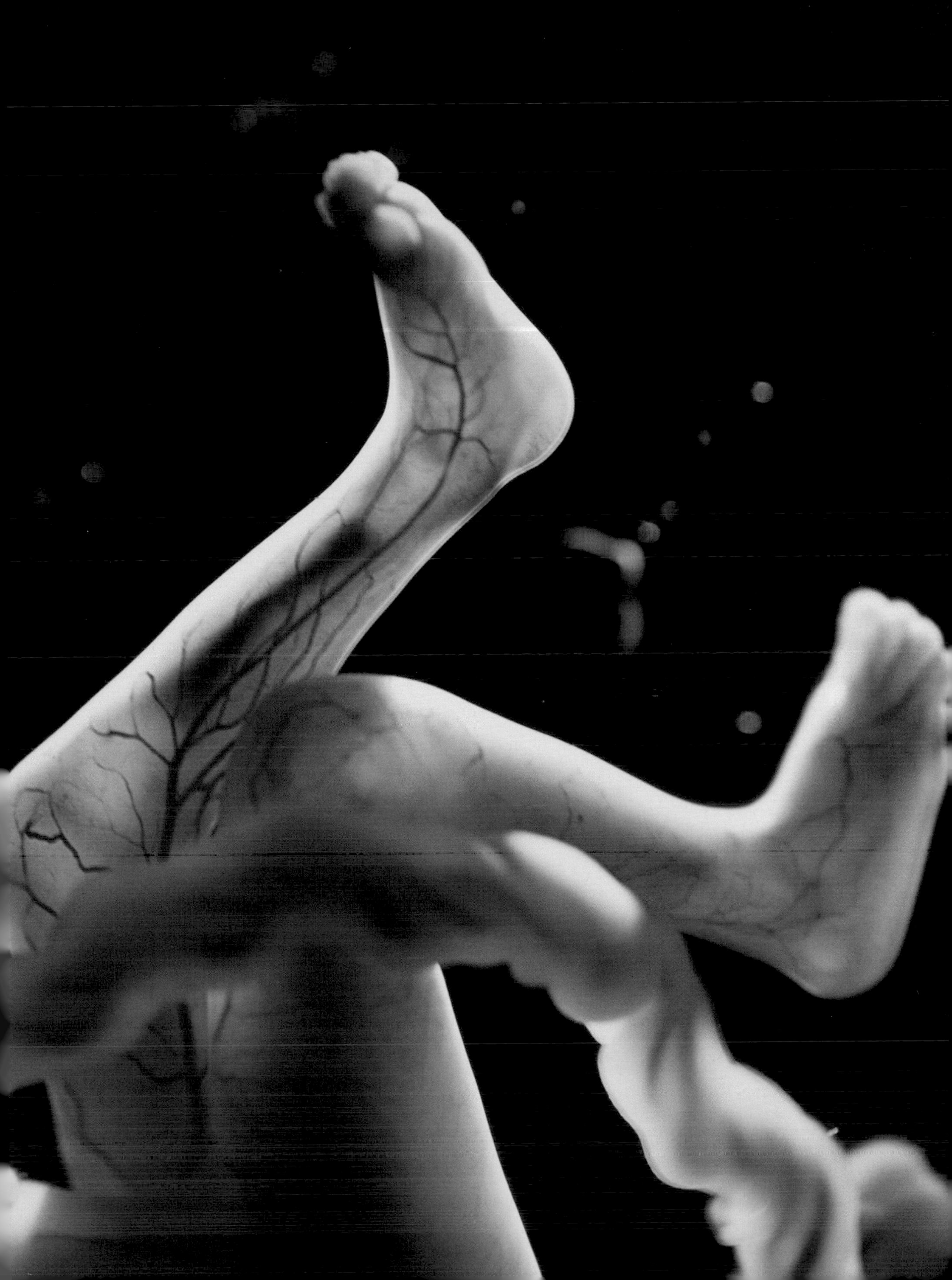

Proper exercise can help relieve discomfort, including back pain.

Keeping in good physical shape during pregnancy is very important, and if a woman has a sedentary job, it is especially important to get frequent physical exercise. Ordinary brisk walking is a good start. A woman with a regular fitness routine can usually continue it in the first part of her pregnancy. As it progresses, however, she will find it necessary to alter her exercise patterns and perhaps even the kind of exercise she chooses.

A woman can strengthen the muscles of her pelvic floor by learning to tighten them. At least during a first pregnancy she will probably need professional guidance to learn the right ways to do this. It is important to work with these muscles regularly both before and after delivery, so that the muscles that have been under so much strain can regain their original resilience after the baby is born. There is also evidence that pelvic floor training helps women experience less tearing during childbirth. After childbirth, some slight "stress incontinence" is quite common, especially when coughing or laughing; these exercises can help solve that problem as well.

In the first three months of pregnancy, the woman likely gains only a few pounds. The ideal amount of weight to gain by delivery depends on the woman's BMI (Body Mass Index) at the start of pregnancy. About half this added weight is accounted for by the baby, the placenta, and the amniotic fluid. Many women gain much more, and a few a little less. In early pregnancy a woman may lose weight due to nausea. In the last few months many women gain a lot of weight because they are retaining water. This happens because the circulatory and lymphatic systems lose their ability to manage water properly, but the normal balance is regained relatively soon after delivery.

APPROX. WEIGHT GAIN
by the end of pregnancy
attributable to:

Baby	3.4 kg/7.5 lbs
Placenta	0.7 kg/1.5 lbs
Amniotic fluid	1.0 kg/2.2 lbs
Uterine lining	1.0 kg/2.2 lbs
Breast size	1.0 kg/2.2 lbs
Mother's circulatory system	1.4 kg/3 lbs
Increased fluids in mother's tissue	1.5 kg/3.3 lbs
Mother's lipid depots	3.0 kg/6.6 lbs
Total	13.0 kg/28.6 lb

The middle trimester of pregnancy is a positive time for most expectant mothers, but minor problems do often occur as well. Low energy levels and tiredness may result from iron deficiency, as pregnancy consumes a great deal of a woman's iron reserves. Her own blood supply and that of the fetus are increasing, and iron is needed for this buildup. For this reason supplementary iron is often recommended early in pregnancy, but in reasonable amounts, as high doses of iron tend to cause constipation, which may be a problem in its own right.

Increased vaginal discharges are common during this part of pregnancy. Sometimes they are caused by fungal infections; from a medical viewpoint they are seldom cause for concern. Most fungal infections can easily be cured with medication. Occasionally discharges arise from a bacterial infection; these must be treated because they may have a negative impact on the cervix and cause early onset of labor.

In this phase of pregnancy some women feel short of breath. Although this sensation becomes more frequent later in pregnancy, a sudden exertion may bring it on now. Many pregnant women complain of backaches. Pregnancy is a great strain on the back, as increasing belly size has to be balanced by backward-leaning physical posture. Many good exercises can counteract back strain and should be started early in pregnancy.

A pregnant woman should bring up any problems or concerns at her prenatal visits, when her blood pressure and her blood count are taken, her weight is checked, and if necessary, protein or sugar in her urine is tested.

A woman carrying twins must anticipate that her abdomen will grow more quickly and that she will have an increased risk of developing complications. Twin pregnancies are referred to as high risk both for mother and children. The developing twins must compete for space in the womb, and sometimes also for oxygen and nutrients. The mother usually has to be examined frequently, with repeated ultrasound scans and blood flow analyses, so that everyone involved will have a good idea of how both babies are faring throughout the pregnancy.

Special prenatal exercise classes are often available in which the movements are tailored for pregnant women, including prenatal yoga, water aerobics, and gym classes. Some of these exercises focus on training the muscle groups that will be needed during the delivery, such as the pelvic floor muscles.

Life in the womb

In the womb, the fetus is sheltered and safe. The temperature is always just right, and the supply of nutrients is more or less constant. By midpregnancy the fetus has begun to explore its own body and environment using its hands. It often holds on to the umbilical cord, and when a thumb approaches its mouth, it will turn and begin to make sucking motions with its lips—a survival reflex. The baby must be able to grip and suck immediately after birth, and pull itself up to the mother's breast. So the fetus is in constant practice, kicking its legs and waving its arms. With every passing day the fetus becomes stronger and more agile.

The fetus is also using its sense of hearing for orientation. Although the ear is structured early, it cannot perceive sound until weeks 18 to 20. The womb is not a silent world, and once the fetus's hearing begins to work, it can hear the mother's digestive system gurgling, her heart beating, and her blood vessels streaming. Her voice imprints early into the consciousness of the fetus. Other sounds probably play no role as yet, but toward the end of the pregnancy external sounds penetrate, and many fetuses react, for instance, to music. The fetus may, like the mother, have a favorite melody. If the parents talk and sing to their baby now, the baby will recognize and prefer their

Week 19. A fetus sucks its thumb, rocked to sleep in the protective cocoon of the amniotic sac. Thumb-sucking is not a form of comfort, but a way of practicing a reflex in utero that will be important for survival after birth.

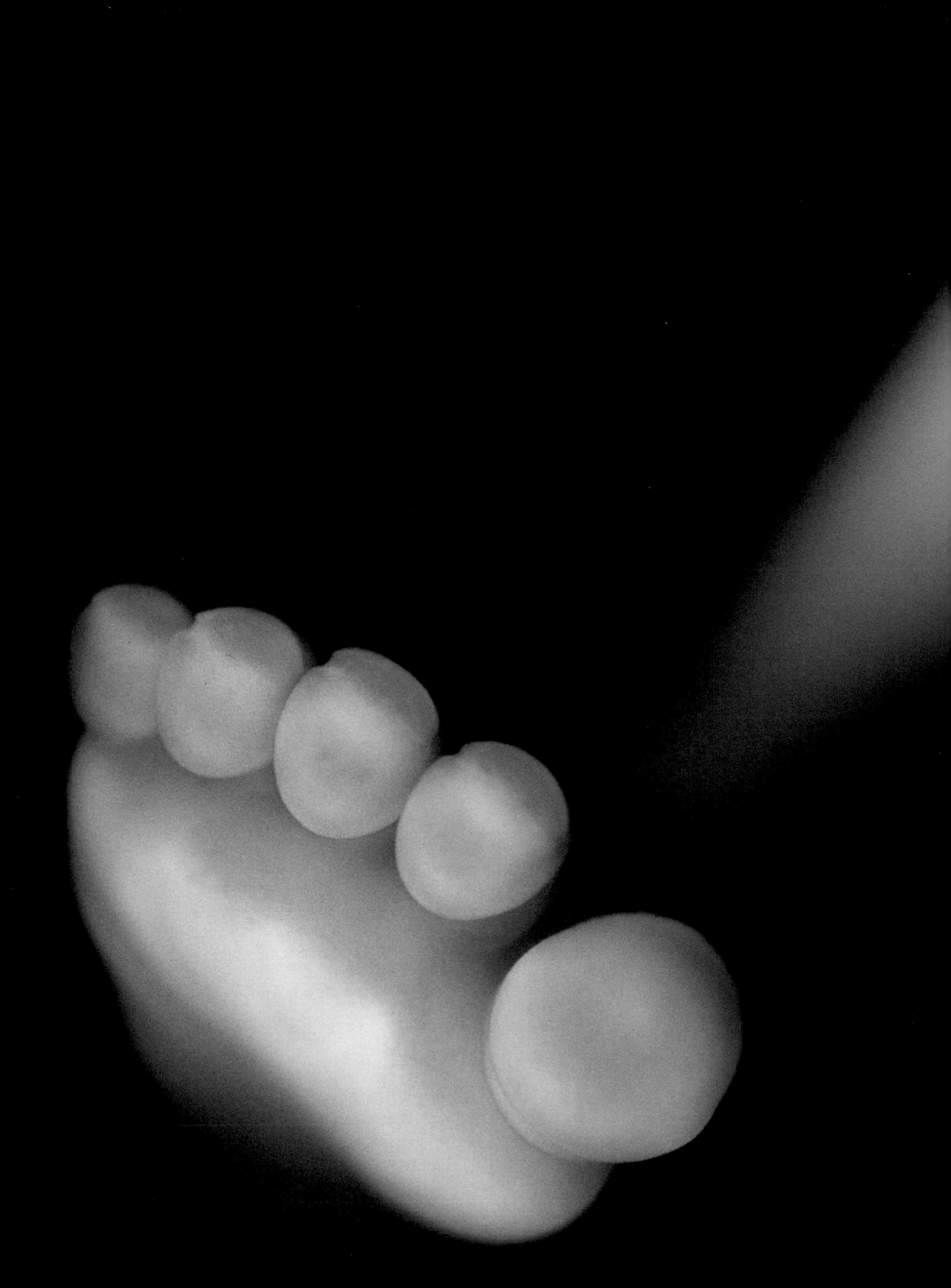

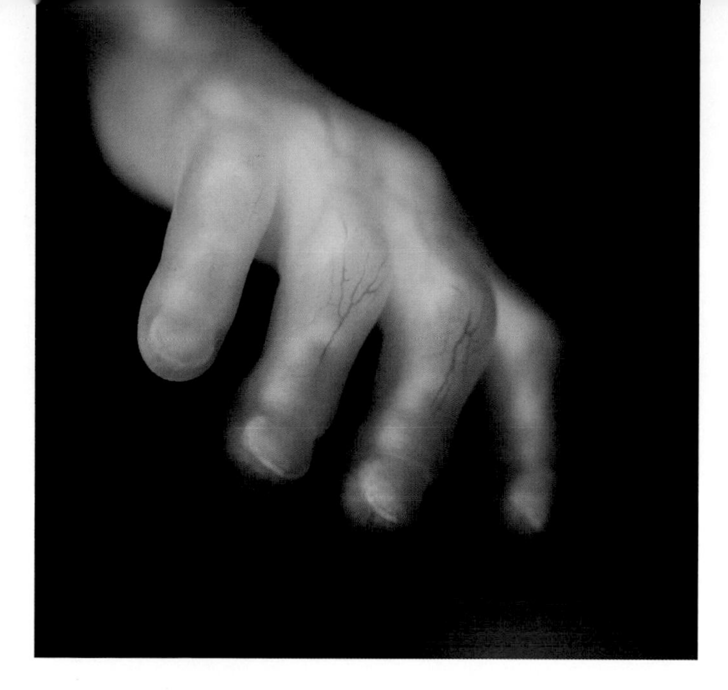

Week 19. The nail beds are clearly visible on the fingers.

voices after birth. Loud sounds may both stimulate the fetus and expose it to stress, quickening its pulse considerably.

Although the eyes remain shut until week 26, the thin eyelids do not close out light completely. It is known that the eye can sense light as early as the third month of pregnancy. Sometimes when an endoscope is inserted into the amniotic sac, a fetus tries to protect its eyes from the light on the instrument, either by turning away or by using its hands and fingers. Light impulses from the eye pass along relatively long nerve pathways to the vision center of the brain, which is at the very back, near the neck. As the brain and the retina develop and become coordinated, the vision center becomes able to classify impressions as light or dark, to sense nuances of color and shape, and gradually to structure complete, coherent visual impressions.

Just halfway into the pregnancy, the hands have already learned to grip.

< At week 19 the feet and toes are quite developed. The fetus moves constantly, kicking against the walls of the uterus.

Beginning early in development the fetus swallows amniotic fluid, which passes through the mucous membranes of its nose and mouth. During its time in the womb, it develops its senses of taste and smell. As a result, a newborn immediately reacts positively or negatively to tastes that are sweet, salty, or bitter, either on the mother's nipple or in breast milk.

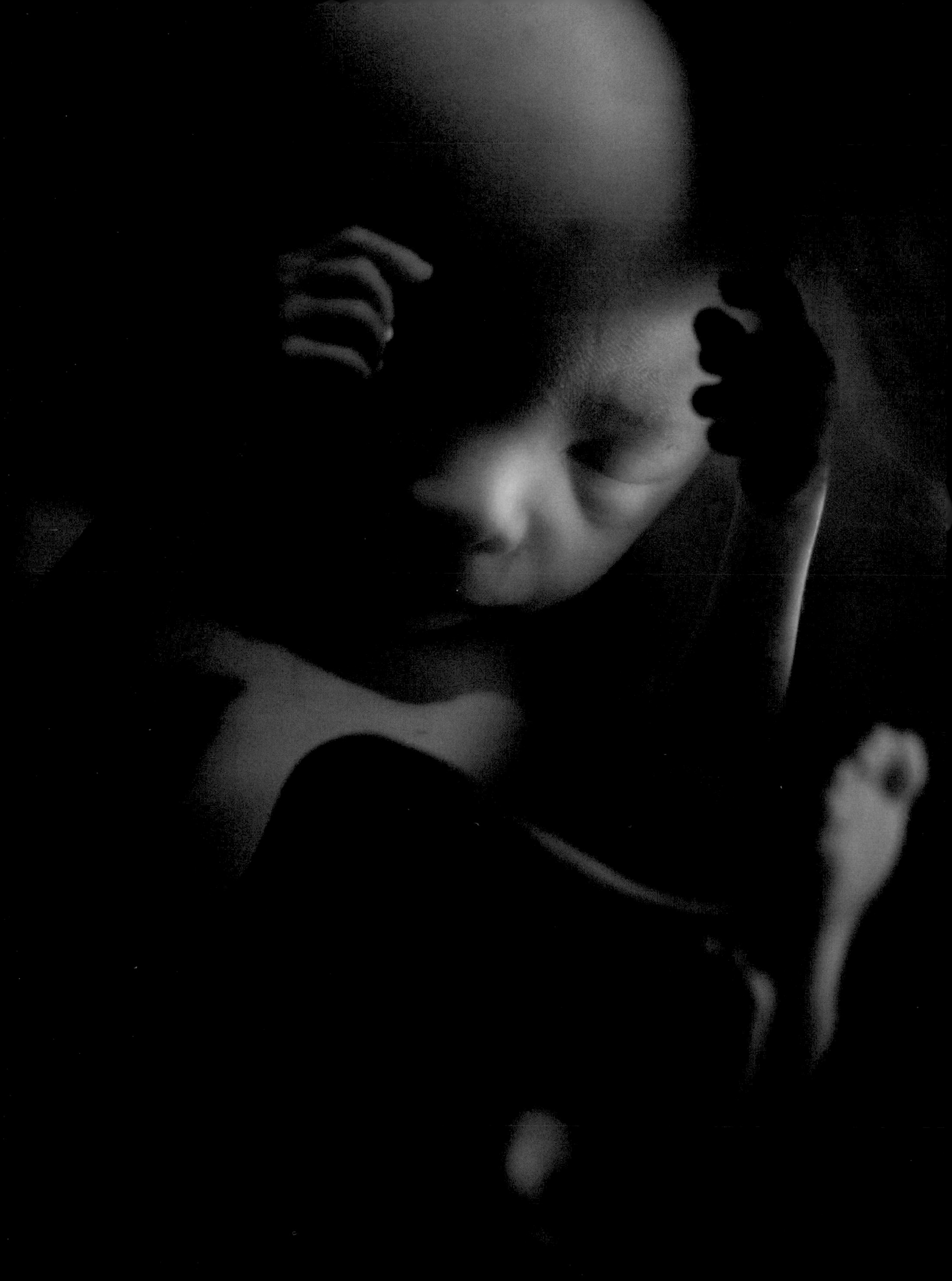

Almost ready for life outside

Two-thirds of the pregnancy is now past. The uterus is still quite a roomy environment for the baby, and in most cases, the comfortable, health-inducing soak in the amniotic fluid will continue for a few more months. The long umbilical cord can twist around the body without impinging on the baby's freedom of movement. The baby can still change position and busily practices various movements. Soon, however, conditions will seem cramped, and many babies then appear to prefer lying head downward. Usually by week 36 the baby has decided whether to be born head or feet first.

The baby is still slim and has not yet accumulated fat under its skin, which is red and thin. Not until the last four weeks of pregnancy does the baby really fill out to fit its skin and develop that plump newborn appearance. Still, weight gain is continual, about 200 grams (7 ounces) per week.

Now the brain develops very rapidly. The cerebral cortex becomes furrowed and convoluted to make room for all the nerve cells. The organs, including the lungs, are maturing steadily, and breathing practice begins to play an important role. Sometimes breathing practice results in hiccups, experienced by the expectant mother as rhythmic abdominal twitches. When the baby's feet touch the uterine wall, its walking and crawling reflexes are triggered.

When the mother-to-be enters week 27 of pregnancy, the baby's eyes and eyelids are so well developed that the eyes are now open and blinking at regular intervals. If the eyes open later, the fetus may have been exposed to something harmful during development. Such damage is most often alcohol-related. Hearing is developed and refined during this period, and the baby likely perceives more and more sounds from the outside world.

The baby now begins to sleep for longer stretches at a time. It often sleeps more late at night and during the mornings, and is more active in the afternoon. Mothers often observe the same pattern after the baby is born. Inside the womb, the baby may sleep for 40 to 60 minutes at a stretch. Newborns, too, usually sleep in short bursts—something the expectant parents may want to plan for.

In week 26 the baby is about 32 centimeters (12.6 inches) long and weighs nearly a kilogram (or about 2 pounds). The baby's face is completely developed except that the cheeks are not yet filled out. This means that the eyes, which now open and close, seem to protrude. During the remaining weeks in utero, the baby will gain a couple of kilos (or about 4 pounds). Lifting the arms and hands at the same time is part of a reflex, governed by the nerves. The same movement can be observed in a newborn.

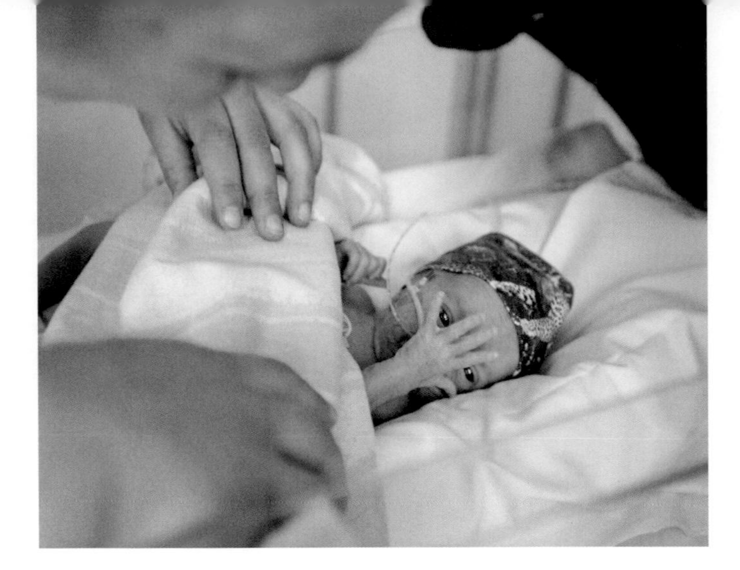

A four-day-old baby, born at 32 weeks and 2 days, weighing 1,690 g (3.75 lb).

Born early

Forty weeks after the first day of the last menstrual period is considered the length of an average pregnancy. But only slightly more than half this time in the uterus is absolutely needed if the fetus is to survive.

In Sweden, about one in twenty children is born prematurely, defined as before the beginning of week 37. In the United States the corresponding figure is more than one in ten. Not many years ago a delivery prior to three months before the due date was a disaster. Today, at least in the developed countries, 82 percent of the babies born in or after week 25 survive. But the beginning of life—just like the end of life—is a difficult tightrope walk in terms of medical ethics. A very premature baby may be able to survive, but sometimes with grave vision or hearing impairments, or intellectual disabilities (if it proves impossible to protect the immature brain). How premature is too premature? Staff at intensive care units of neonatal wards must constantly consider and reconsider this question.

In recent years new understanding about the condition of the lungs in the fetal stage has made it possible to help even very small babies survive. Today there are babies who have survived and developed normally who were born as early as week 23 and weighed less than 500 grams (1 pound). But immature lungs can cause irreversible brain damage owing to impaired oxygen supply. The lungs develop much later than, for example, the heart, which supplies

This baby was delivered by cesarean section when its mother developed preeclampsia. Premature babies receive the care they require in the neonatal unit.

> Just a few decades ago, this child would probably not have survived. Now she was kept alive and survived with no harm done, thanks to modern technology. Developmentally, she is expected gradually to catch up with babies carried to term.

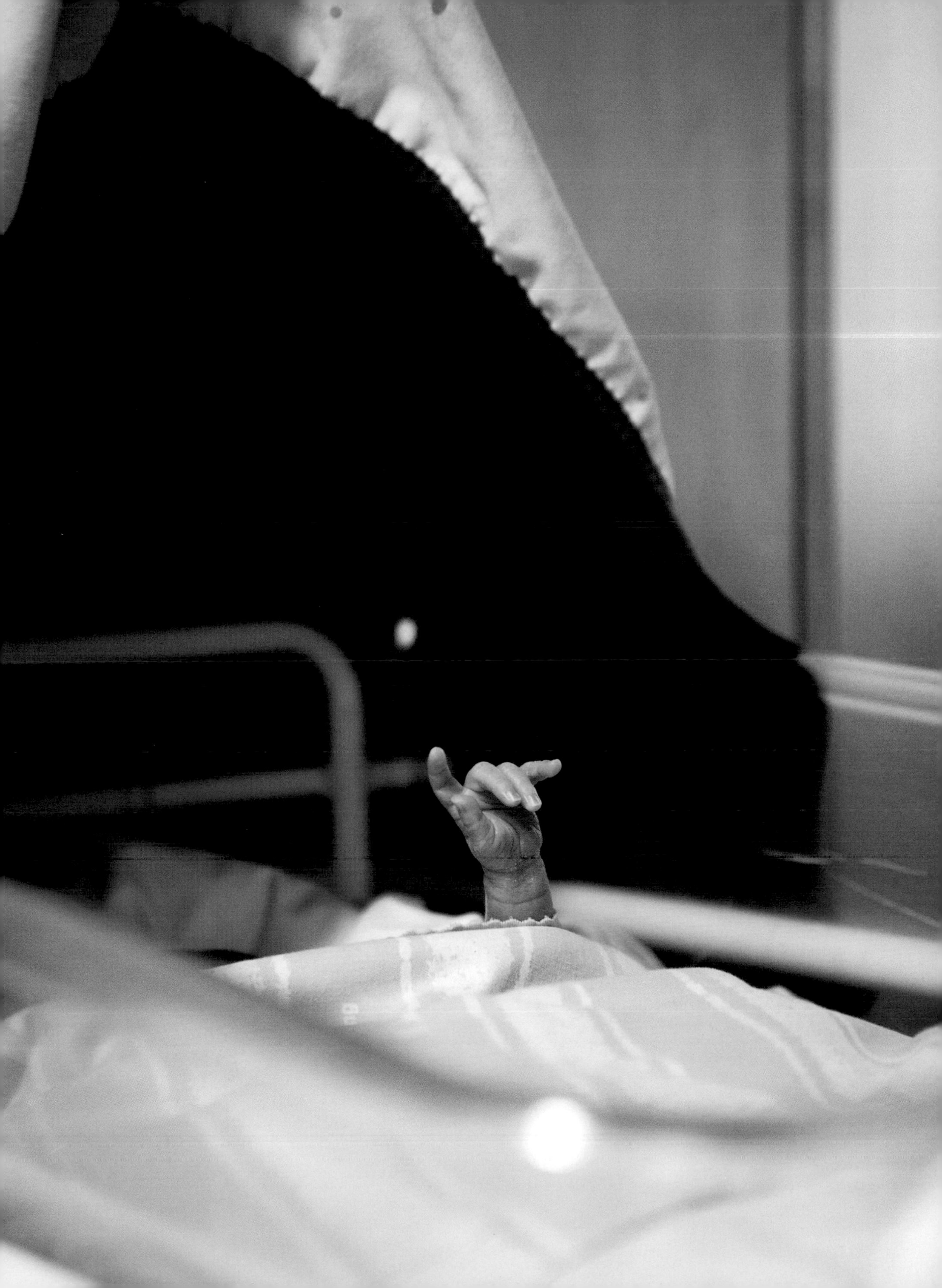

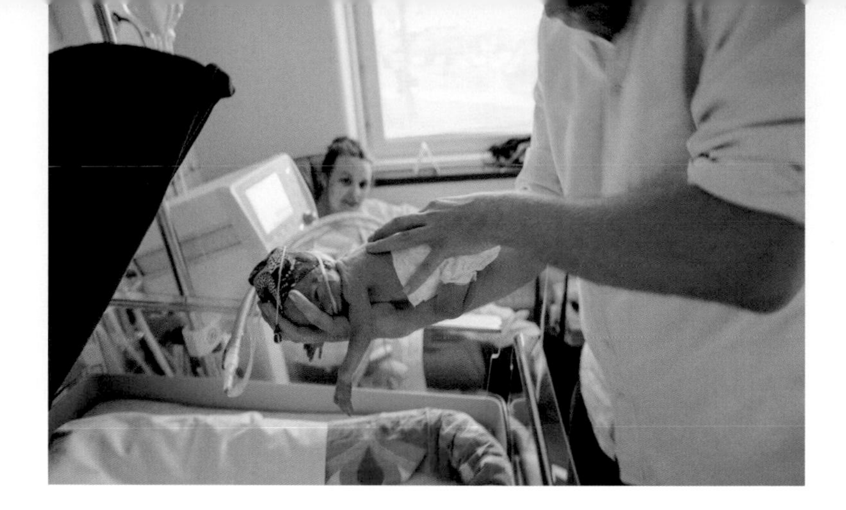

Both parents typically help care for their baby in the neonatal unit.

oxygen from the placenta to all the organs in utero. Because the placenta transfers oxygen from the mother, and the oxygen supply travels to the fetus via the umbilical cord, the lungs serve no real function in the womb. Not until the moment of birth, when the oxygen supplied by the umbilical cord is abruptly terminated, are the lungs really needed.

The immature lung is subject to a condition called respiratory distress syndrome (RDS). RDS occurs when the lungs do not produce a lubricant (surfactant) needed for the air sacs to open and stay open, so as to transfer oxygen. The sacs also contain fluid. This is why a great deal of the life-saving effort that goes into helping babies born prematurely is focused on making the lungs sufficiently mature so that the baby will be able to breathe independently.

Today an obstetrician has several ways to prevent the delivery from beginning too early and to help an immature fetus remain longer in the womb. Bed rest for the mother is important, but there are also drugs that can inhibit the onset of labor and let the uterus relax. As always in pregnancy, regular checkups are essential, particularly if there is any reason to anticipate the problem of early contractions. Postponing delivery by as little as a couple of weeks may be what is needed to allow the lungs to mature, making it easier for the baby to breathe and supply all the organs with oxygen. Treatments with the hormone cortisone speed up lung maturation. Even if the amniotic fluid has leaked out at an early stage of pregnancy through a hole in the amniotic sac, much can be done. Amniotic fluid is continually produced, and with bed rest and careful monitoring, labor may be postponed by several valuable weeks. But after rupture of the membranes, the risk of vaginal bacteria making their way into the uterus increases, sometimes necessitating antibiotics, and delivery.

It is particularly important for premature babies to get off to a serene, secure start with lots of close contact.

< Both parents can cuddle the baby on their chest to provide security and warmth. This baby, whose lungs are not yet fully developed, is receiving help to breathe.

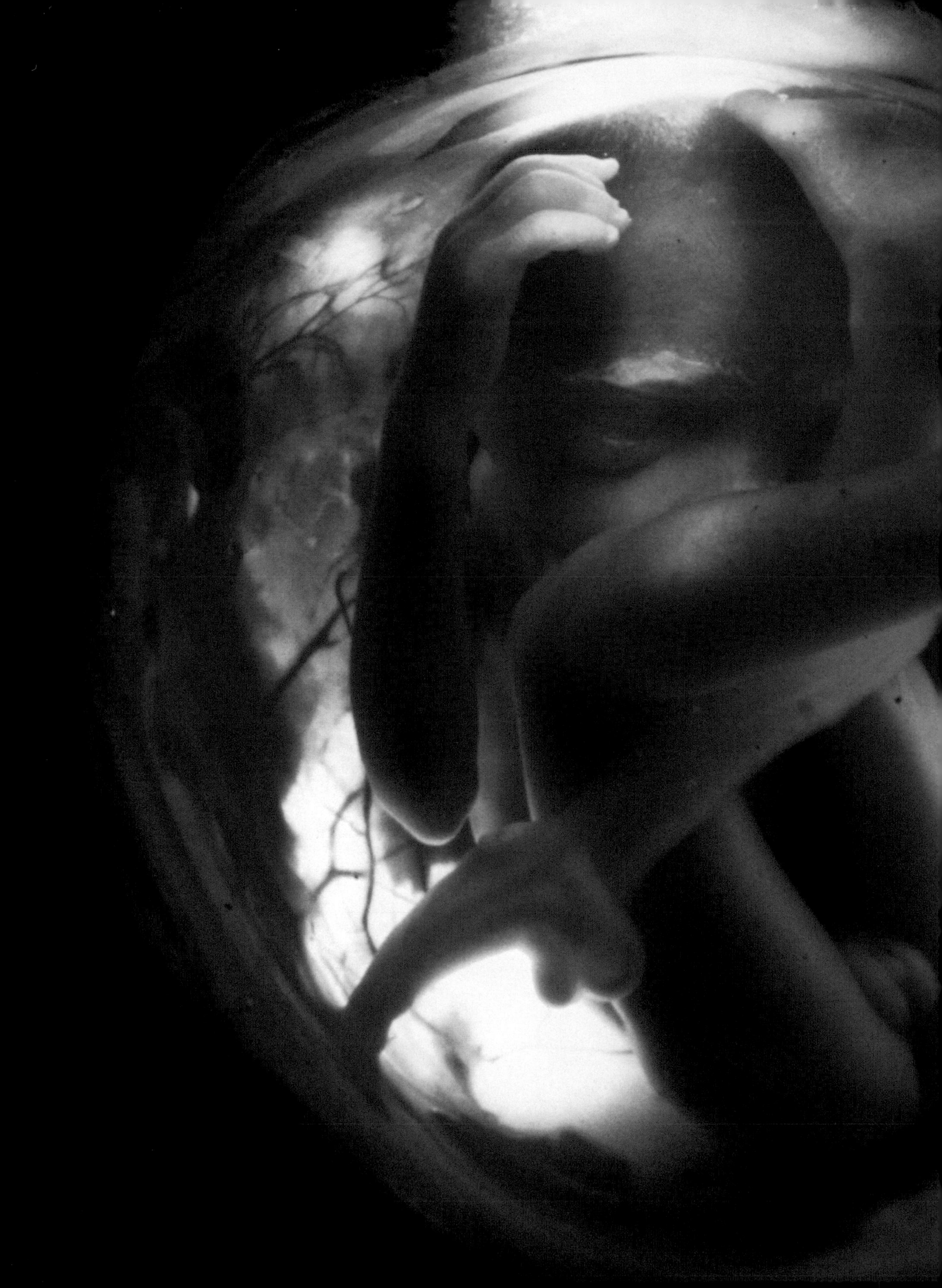

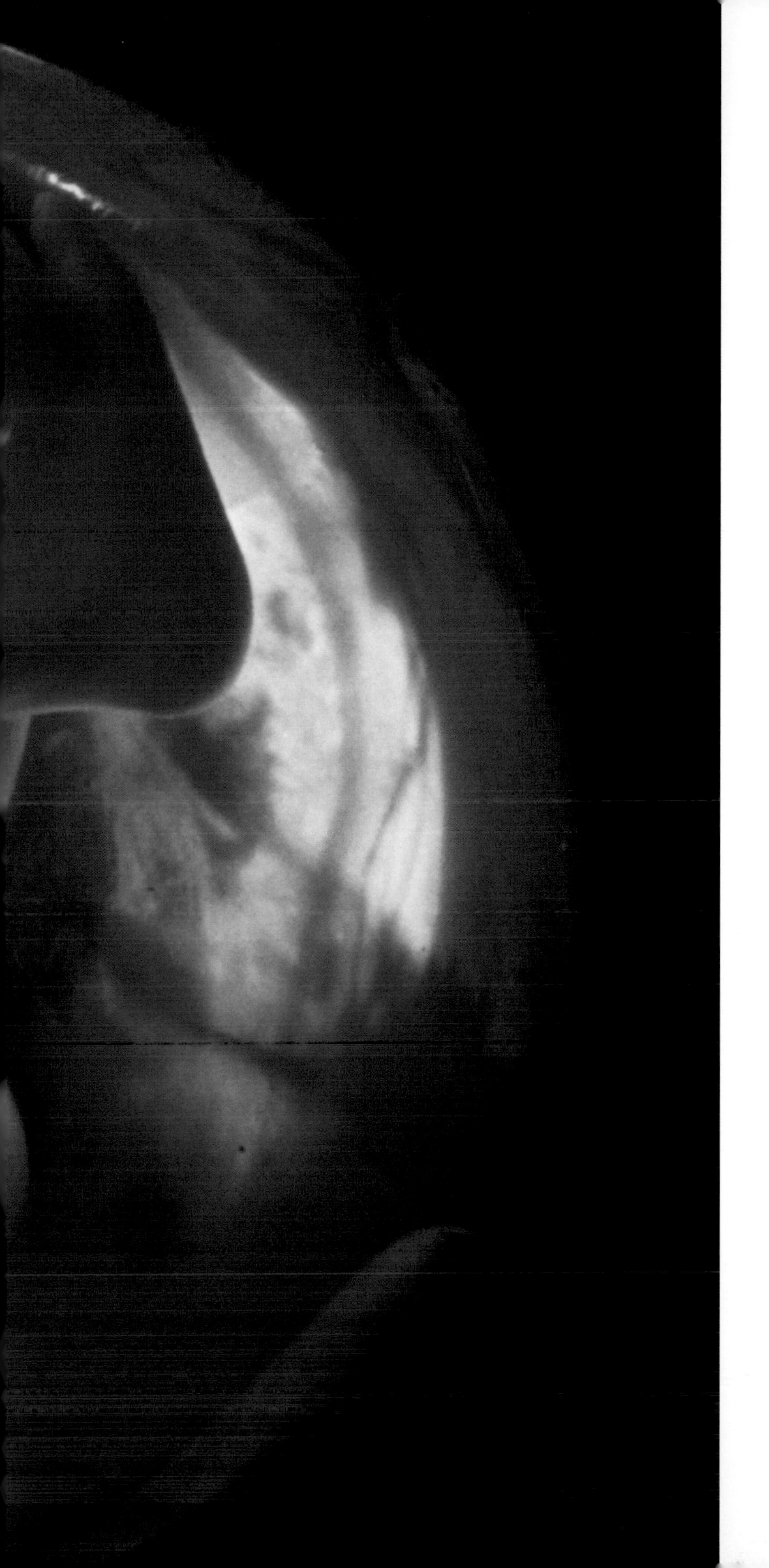

Week 28. The baby can still change position in the uterus. It busily practices various movements later displayed by the newborn. The blood vessels of the extremely supple umbilical cord are embedded in a gelatinous mass.

Getting to be a heavy load

Most pregnant women have felt very well for several months, and pregnancy has seemed easy and fun. Now, as the end approaches, new little health problems may develop, and sometimes larger ones as well. Acid indigestion and heartburn are common, especially toward the end of pregnancy; liquid medications and pills can provide effective relief without entering the mother's bloodstream, so there is no risk of hurting the baby.

Leg cramps due to circulatory difficulties are a common problem in the second half of pregnancy. A pain that radiates from the back down into the legs may cause severe discomfort in late pregnancy.

Around half of all pregnant mothers will experience a condition called *diastasis symphysis pubis.* In some cases the pain can be debilitating, but in others it causes only minor discomfort in direct relation to how much strain is placed on the body. The cause is hormonal, and it results in a separation of the right and left pubic bones due to a softening of the ligaments that hold them together. Diastasis makes the pelvis unstable and makes it difficult to walk. The joint may become extremely sensitive to pressure. Though painful, the separation actually helps with delivery, as it increases the elasticity of the birth canal. It is important not to place undue strain on the pelvis, for example by going up and down stairs or performing heavy lifts. Rest is also very important to reduce stress on the joint. For this reason, in Sweden, women

Even women who sail through early pregnancy may experience fatigue and discomfort as the end draws near. Most can maintain their daily routines, but relaxing and avoiding undue stress become even more important.

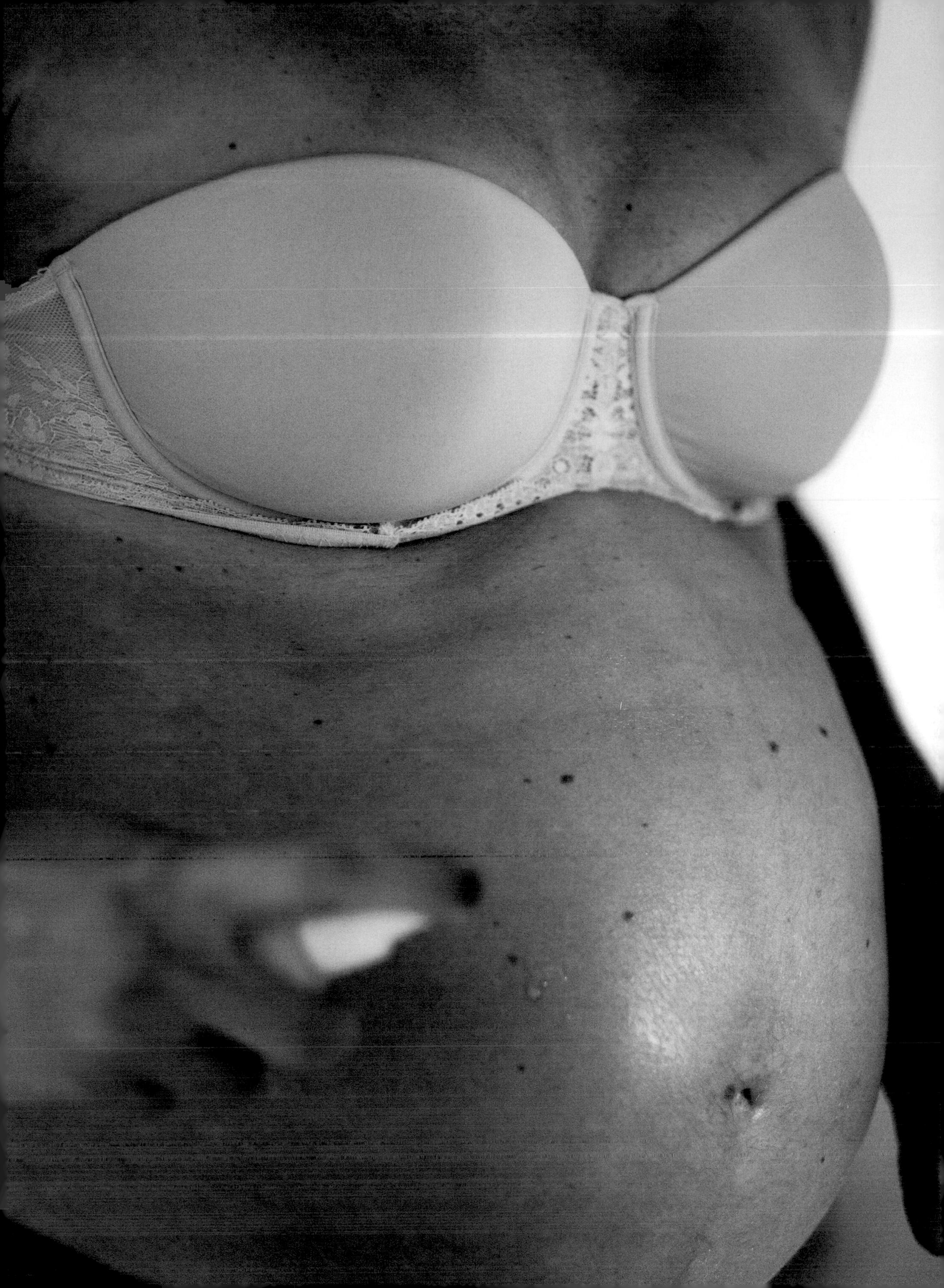

whose jobs involve heavy lifting are eligible for a maternity allowance beginning in week 32. It often feels as if the condition has vanished immediately after delivery, but this is rarely the case. Recovery time varies, and in some cases it may take quite a while.

Varicose veins and hemorrhoids are also frequent and sometimes very aggravating complaints. Both are caused by the pressure of the enlarged uterus on the major veins that run through the pelvis. Blood has difficulty passing back through and therefore collects in the veins below the pelvis, which become swollen. This can result, in the long run, in damage to the valve system of the veins, the function of which is to prevent the blood from flowing backward. Severe varicose veins do not disappear on their own, but as a rule they are not operated on until at least a year after the delivery. Most doctors also recommend postponing an operation if the mother is planning for more children, as the varicose veins will return during the next pregnancy. Hemorrhoids seldom require surgery and usually heal by themselves fairly soon after delivery, especially if any constipation experienced by the mother during pregnancy clears up, as this can be a contributing factor.

Most women retain some fluid and experience some swelling during pregnancy. Fluid retention is especially common in the feet, ankles, and hands late in pregnancy. Moderate bloating is not a cause for concern.

The mother-to-be should try to avoid major physical and emotional stress at this stage of her pregnancy. Intense stress may have negative effects on both mother and baby and is one common reason for women being given sick leave toward the end of pregnancy. Of course, each pregnant woman must evaluate her own work situation and check with her physician in regard to any unusual stressors in her work environment.

An expectant mother with a history of back pain is especially likely to experience back pain during pregnancy, as she is forced to balance her growing belly with a backward-leaning posture.

A great test of strength—the delivery—is fast approaching. Time to take things a little easier.

Only a few weeks left

Until week 37, the baby has plenty of room to move and even turn a somersault. But before long, space in the uterus is tight indeed. The volume of amniotic fluid decreases a little after week 35. At delivery there are usually about 0.5–1.5 liters (1–3 pints) remaining.

During the final month in utero, the baby gains about a kilogram (2.2 pounds). If the baby should be born at this point, survival is not a problem, and statistically speaking, most twins are nearing the moment of birth.

Most babies (95–97 percent) are now upside down, with their heads quite a way into the pelvic canal of the mother. She becomes aware of this when she notices that her belly is now lower and finds it easier to breathe. When the baby's head has lodged in the pelvic canal, it is said to be engaged.

Some babies seem to decide early on not to lie head down; in weeks 36–37 an attempt will often be made at the hospital to turn such babies around. After an ultrasound exam, the mother will be given medicine to relax the uterine muscles and make more space for the baby. The doctor will massage the mother's abdomen, applying external pressure to direct the head downward. If this attempt does not succeed, a cesarean delivery will usually be planned.

A few weeks before birth, the baby gets in position, usually with head lodged in the pelvic canal. If the baby cannot be coaxed into this optimal position, a cesarean may be planned.

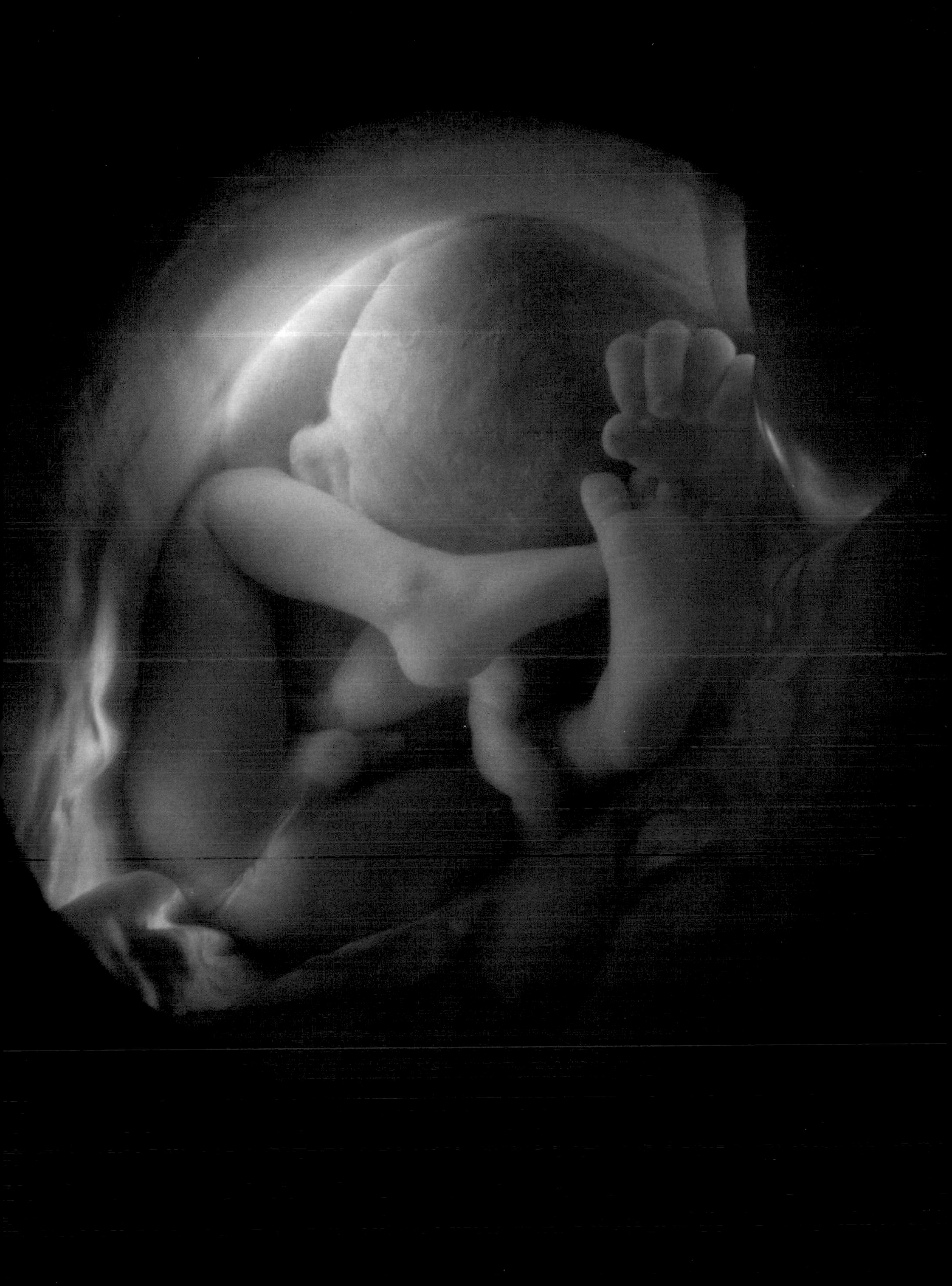

Delivery time approaches

When there are only a few weeks left, time seems to drag. The baby has lodged in the pelvis, making the woman's abdomen somewhat less cumbersome and breathing slightly easier. Tingly legs are now a matter of course; the baby may kick or punch hard. The enlarging uterus puts pressure on and compresses the mother's bladder, causing her to rush to the bathroom frequently. Increased sensitivity to heat and some swelling—particularly of the feet and fingers as the day progresses—are very common. The mother sweats more and it is difficult to sleep soundly with a belly that seems in the way no matter in what position. Today mothers are advised to sleep on their left side during the final weeks of pregnancy.

Beginning in week 29, the mother-to-be will usually go to prenatal appointments every other week. The doctor or midwife will check her blood pressure and may take a urine sample. They will also listen to the baby's heart and measure the size of the mother's abdomen. These measurements will be plotted against a normal curve to make sure the baby's growth rate seems normal.

The appointments are probably reassuring as well as necessary at this time. An expectant mother will have lots of questions. How long do I have to wait? Can't labor be induced on my due date if nothing has happened by then? Do the risks increase for a baby not born on time?

It is the job of the doctor or midwife to help the parents prepare for the delivery by answering any questions they may have. It is important that the parents trust in their own ability to help make the delivery go smoothly. A midwife or doctor can teach the expectant mother relaxation techniques and suggest ways that her partner can support her during the delivery. The parents-to-be may also attend prenatal courses, such as Lamaze classes.

Parents also need to learn what a normal delivery is like, how breastfeeding works, and what to expect in the first days and weeks after birth. It is also a good idea to prepare for the delivery not running as planned. Any mother may have to be delivered by emergency cesarean section, for example, even when everything looked fine for a vaginal delivery.

Most women have occasional contractions late in pregnancy. For a moment the uterus becomes hard as a ball, only to relax again. These painless contractions, also known as Braxton-Hicks contractions, are more common in women who have already given birth than in first-time mothers.

(overleaf) It is certainly an advantage if both parents can attend prenatal appointments and take any prenatal courses together.

WILEY

Some expectant parents are able to visit the hospital maternity unit or birth center where their baby will be born; generally this occurs a month or two prior to the due date. Knowing as much as possible before delivery helps put most parents at ease. For first-time parents, this may also be the moment when they fully realize: "We're going to be parents, and this is where our baby will be born!"

In many countries today, it is considered natural for the parents to be together when their child is born. For most mothers, having their partner in the delivery room is a great emotional and physical support. It can also endow the couple with a strong sense of togetherness that can strengthen their own bond and help get their lives as parents off to a good start. Many partners look back on the deliveries of their children as among the most fantastic experiences of their lives. But a delivery can also be stressful, especially if labor is prolonged or complications arise. Some couples therefore engage a professional labor companion, known as a doula, to support them throughout the delivery.

Even if the mother is delivering by cesarean section, the partner will almost always be allowed to attend, except in the case of an emergency cesarean performed under general anesthetic. Today, medical professionals are trained in providing care for the entire family, and they know how important it is for the partner to be involved in the birth. They will help ensure that the parents have a chance to meet their new baby together just as they would after a vaginal delivery. There will be time for the important first moments of bonding and skin-to-skin contact and the first breastfeeding session, in the same way as after a vaginal birth.

Parents expecting twins often feel extra anxiety about the delivery. Even if the mother gets plenty of rest and does everything she is advised to do, twins tend to be born some three weeks early. It is not unusual for the doctor to induce labor in week 38. There is simply no more room in the uterus, and the placentas (usually there are two) cannot go on supplying enough nutrition and oxygen for two. Newborn twins also tend to weigh less than 2.5 kilograms (5.5 pounds) on average, rather than the 3.5 kilograms (7.7 pounds) that is the mean for single births. Risk increases when twins are born vaginally, but a mother expecting twins who has already had a normal delivery will generally be able to give birth to the twins vaginally as well. All twin births are planned and monitored with extra care. In the United States cesareans are increasingly done "just to be on the safe side," but if no unexpected complications arise, a vaginal delivery can be preferable for both mother and babies.

During the final month of pregnancy, the mother will normally have a weekly prenatal appointment. Weighing becomes routine for many.

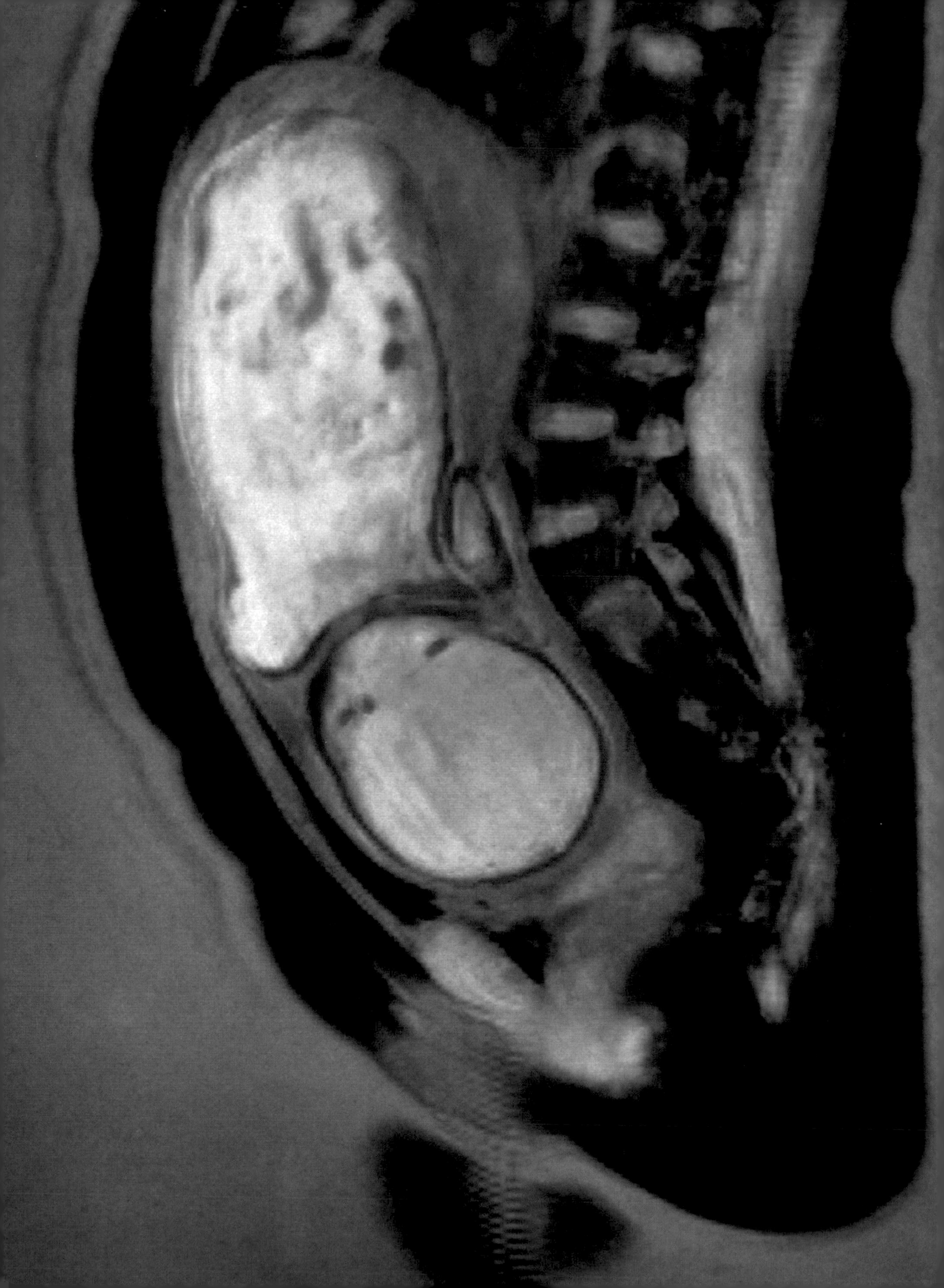

Overdue

Many expectant mothers find going past the estimated due date more trying than almost any other aspect of pregnancy. Their stomach is heavy and uncomfortable, they feel impatient, and may have to field daily queries from anxious grandparents and others. Still, most gynecologists and obstetricians all over the world agree that when it comes to delivering a baby, the wisest thing is to be prudent, because a baby who spends some extra time in the uterus often has a good reason for doing so.

Folk remedies for stimulating labor are many: everything from nipple stimulation to running up and down stairs to castor oil "cocktails." If labor does not begin on its own, it is often initiated before forty-two weeks. After this time the placenta may not be able to provide the growing child with sufficient nutrition. At this point, it is time to induce labor. If the woman's water has broken but contractions have not begun, this is another common reason to induce labor.

Once the decision to induce labor has been made, the woman visits her doctor or midwife. The best method of inducing labor is selected, based upon the "ripeness" of her cervix. Ripeness is assessed on a ten-point scale: the lower the number, the longer it is likely to be before labor begins.

Labor may be induced by either hormonal or mechanical means, or by a combination of the two. The hormones used to induce labor are oxytocin, which stimulates uterine contractions, and various forms of prostaglandins, which soften the cervix and prepare it for labor. Any of several mechanical methods may be used. One method is to insert a balloon catheter into the cervix. Another is known as stripping the membranes. In this procedure, the doctor or midwife sweeps a finger around the upper part of the cervix, separating the amniotic sac from the wall of the uterus. This is thought to stimulate the local release of prostaglandins. The cervix may be encouraged to open at the same time. Finally, if the water has not broken already, the amniotic sac may be punctured. This is called artificial rupture of the membranes. If the attempts to induce labor do not succeed, a cesarean section will be performed. How long to wait before the C-section is decided on a case-by-case basis, by the doctor and parents in consultation.

Sometimes the baby seems to be reluctant to face the world, and the due date passes uneventfully. The head may have descended into the pelvic inlet and stopped there and the cervix is not effaced or dilated. This image was taken using Nuclear Magnetic Resonance (NMR), a technique which unlike X-rays is completely harmless for mother and child.

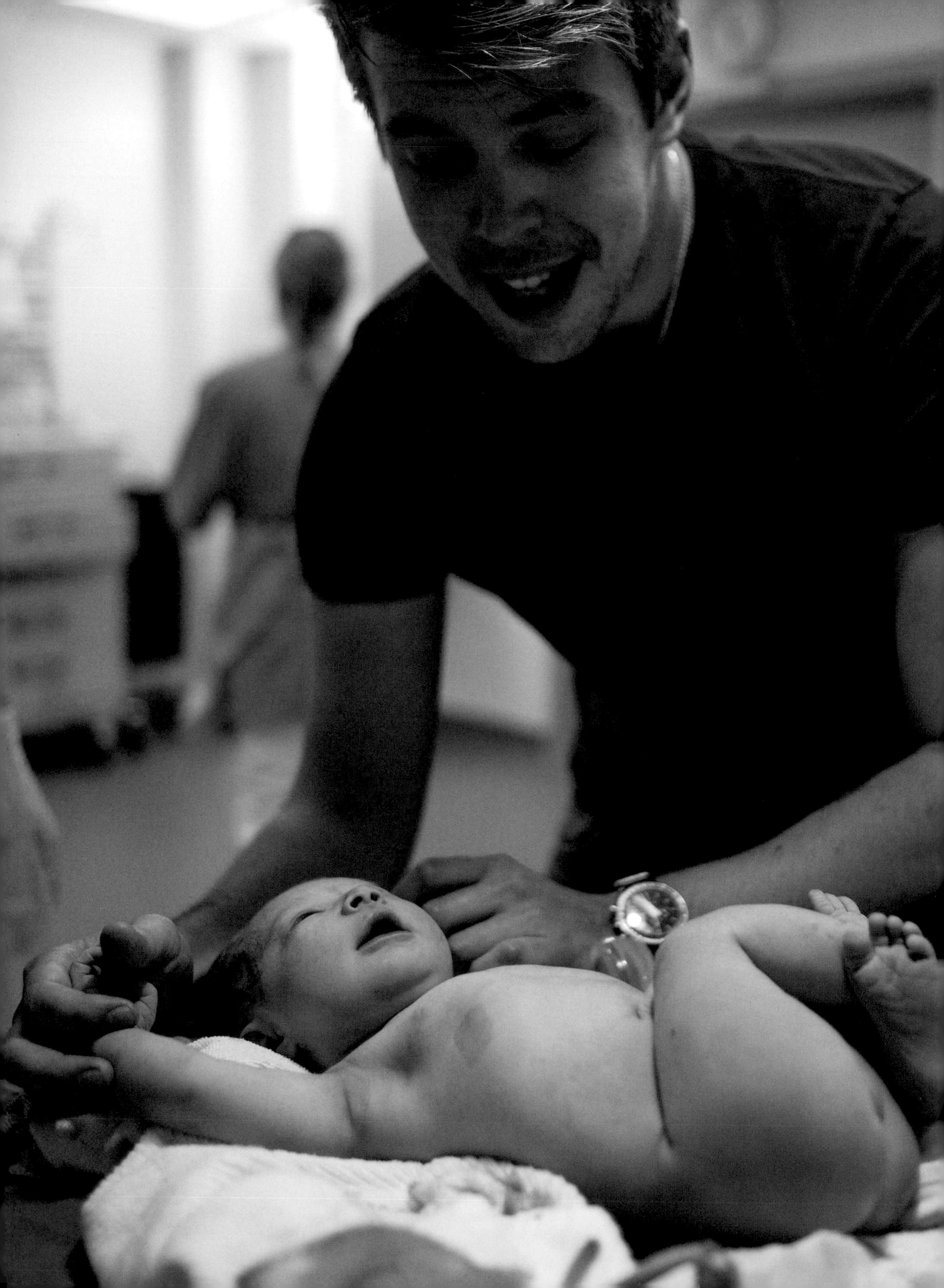

LABOR AND DELIVERY

Nine months of waiting are almost over. Who will the new baby be? Will everything go well? For both parents, the delivery and birth of their child is an intense physical and emotional experience, perhaps the most momentous of their lives. When the umbilical cord is cut and the new baby takes its first independent breath, the world stands still. Each delivery, each birth is unique: an everyday miracle.

First signs

Almost nine months have passed. The due date is approaching, and for the parents-to-be, every day is full of anticipation. The baby might come at any time, but how will they know when? There are three common signs that birth may be near at hand: regular contractions of the uterus, the water breaking, and bloody discharges. Parents should call their caregiver for instructions when any of these occur, although it will not always be time to leave for the hospital right away.

Many women experience practice contractions late in pregnancy. Contractions that become more frequent, returning time after time and with increasing intensity, are often a sign that labor has finally begun. The cervix, normally 3–4 cm (1.18–1.57 inches) in length, now begins to shorten and thin out (efface) and to open (dilate). If everything seems normal, parents are usually advised to leave for the hospital or birth center when the contractions come at four-minute intervals, or at five-minute intervals if the mother has given birth before. Sometimes the contractions stop on the way, and after an obstetrician or a midwife examines the woman, she is sent home to wait until labor begins again.

In about 10–15 percent of women, the first sign that the delivery has begun is that amniotic fluid runs out of the vagina. This is also known as the water breaking. If this happens, it may be time to go to the hospital, where a professional will check the color of the amniotic fluid. If it is green with the contents of the baby's first bowel movement, this may (but not always) be a sign that the baby is no longer faring very well in the uterus. Sometimes the mucus plug in the cervix may also dissolve and be expressed as a heavy discharge, which the mother may mistake for her water breaking. The baby's

Now the mother feels more frequent contractions, perhaps causing a dull ache in her lower back. This is a good moment to call the hospital and ask if it is time to come in.

10B
ALLMÄNNA
IDROTTSKLUBBEN

heartbeat will be measured by an external fetal monitor. The mother will also get a checkup, including taking her temperature to rule out infection. If all seems well, she may be sent home to wait for her contractions to begin. If they have not started after 24–48 hours, a doctor will usually induce labor.

If the first sign of labor is bleeding, the parents should call the hospital or birth center for instructions. A small amount of blood is most common. It can come from the mucous membranes of the vagina or from the cervix as it softens, effaces, and dilates in anticipation of the birth. The mucus plug that has been blocking the cervical canal is usually tinged with blood and may also come loose during the small preparatory contractions. This can occur up to a few days before the actual onset of labor. Intercourse may also cause bleeding. A large amount of blood is an urgent reason to call the hospital and go straight in. Parts of the placenta may have torn loose from the uterine lining, which jeopardizes the baby's oxygen and nutrient supply.

Most babies are born sometime between week 38 and week 42. The beginning of contractions is governed by a number of different factors interacting in the body, particularly hormones. The progesterone that builds up in the placenta during the course of the pregnancy, increasing month by month, has various roles, one of which is to ensure that the muscles in the wall of the uterus remain relaxed and calm. Other hormones present in the body produce an opposite effect, especially oxytocin and prostaglandin. For the birth to begin, the amount of progesterone in the bloodstream has to decrease so that the labor-promoting hormones gain the upper hand, although it is still not known precisely what causes this to happen sometime around week 40.

Most pregnant women have been told precisely when to go to the hospital, but uncertainty often causes many to arrive earlier than is really necessary. Take it easy! There is seldom any terrible rush.

At the hospital

Upon arrival at the hospital, and after a short admissions procedure, the mother has her blood pressure and urine tested. Staff will also check the position of the baby in the uterus and obtain a fetal heart tracing to measure the heart rate of the baby and the frequency of contractions. This exam takes 20–40 minutes, after which time the monitoring devices are usually removed. During the admissions procedure, parents may ask staff about any particular concerns or requests they have. Some parents also prepare a written letter and give it to their midwife or doctor in advance of the delivery.

The next step is usually a vaginal examination, which will give parents and the medical staff a sense of how far labor has progressed. Of course, it is important that the mother herself be allowed to decide when this exam should take place.

The mother is now in the first stage of labor, known as the dilation stage. During its active phase, the contractions become stronger. They last longer, up to 60–90 seconds, become more painful, and occur more and more frequently, until they are about two minutes apart. The cervix dilates at an average rate of 0.5–1 centimeter per hour until it reaches 10 centimeters, when it is said to be fully dilated. Meanwhile, the head of the baby descends through the pelvis, passing the ischial spines (two bony protrusions in the narrowest part of the pelvic cavity) and moving down toward the pelvic floor in a rotating movement. Usually the water breaks toward the end of this stage, after which the contractions often become stronger and more painful.

THE STAGES OF LABOR

Labor stage one, or the dilation stage, begins with a latent phase, which usually lasts 18–20 hours. Most women can relax at home during this phase. The second, active phase begins when two of the following three criteria have been met: the cervix is fully effaced and has dilated one centimeter; the water has broken; or painful contractions have begun. Stage two of labor is the expulsion stage. It is divided into a passive phase, during which the baby descends into the birth canal, and an active phase, when the mother bears down and delivers the baby. The final stage of labor is the delivery of the placenta.

A relaxing massage is one of many ways the mother's partner can make her more comfortable during the first stage of labor.

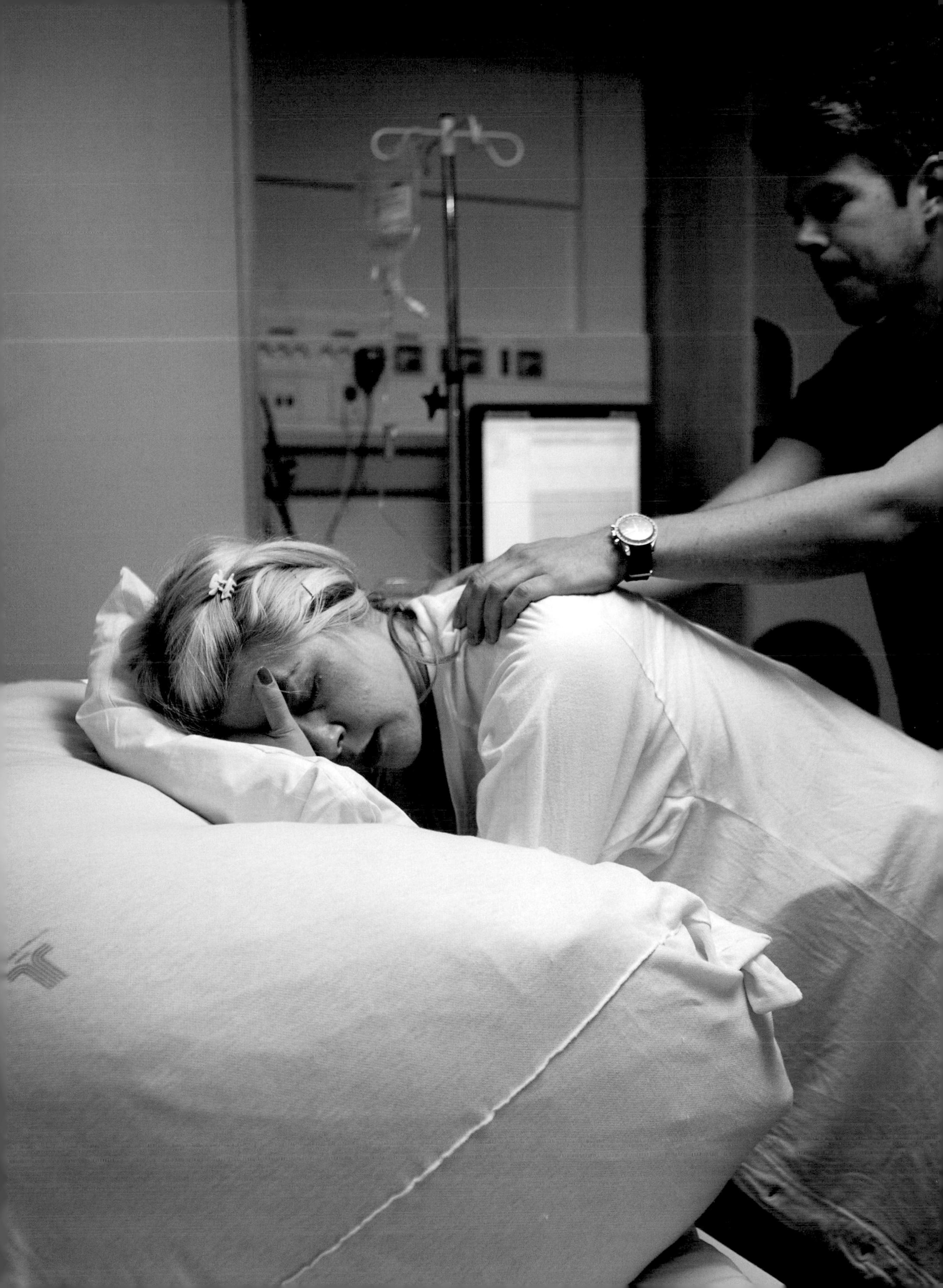

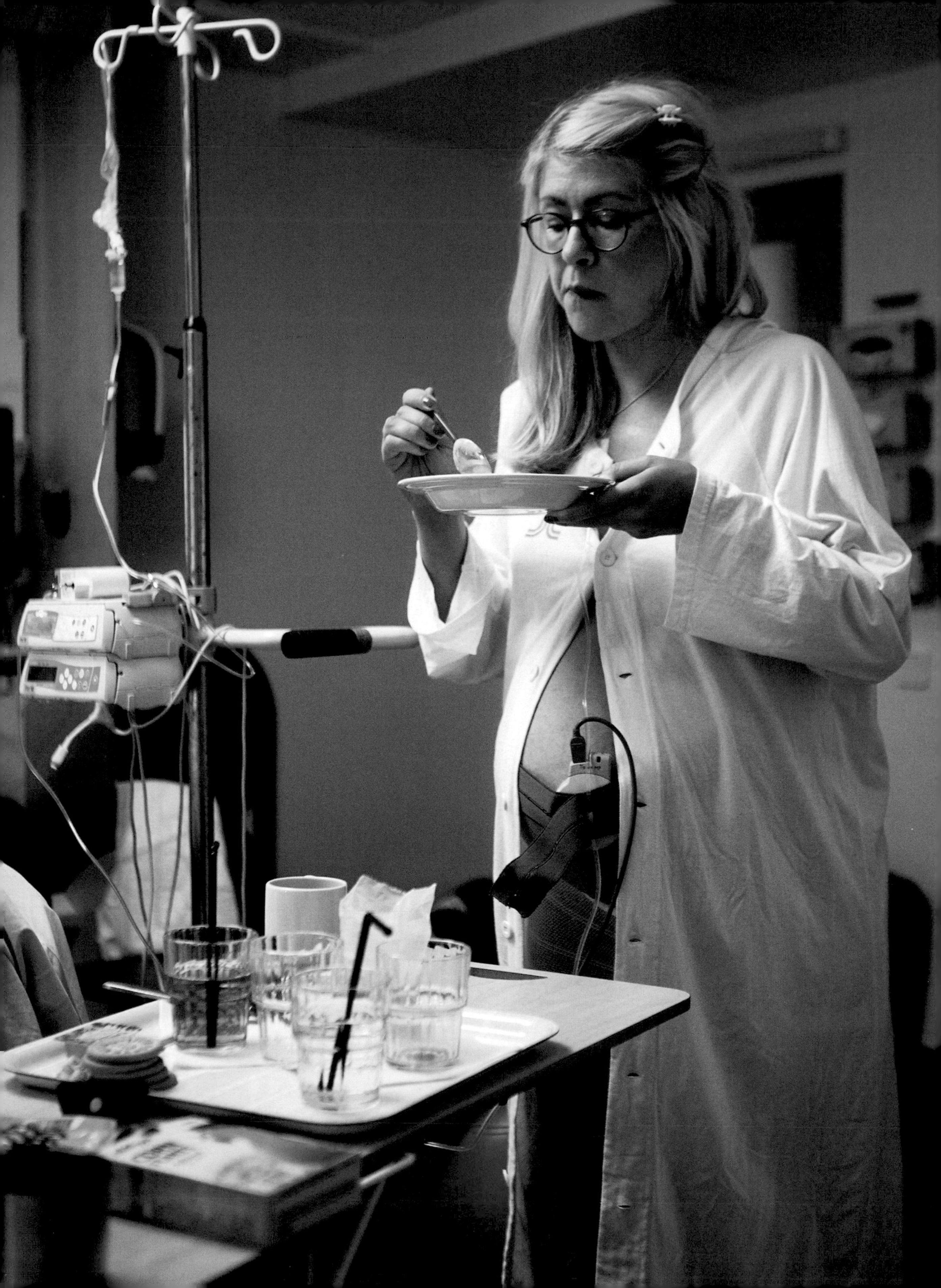

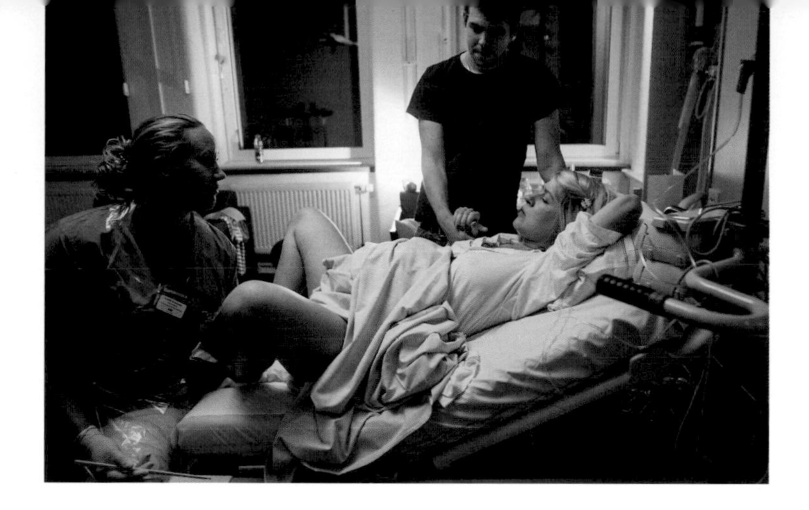

A midwife checks to see how far the baby has descended into the birth canal and how many centimeters the cervix has dilated.

The first stage of labor can be very trying. It is important for the partner and doctor or midwife to offer the mother encouragement and praise. Almost all women will need snacks and drinks to keep their energy up, and physical assistance to change positions. Many women also request a bath or shower, a massage, or help with a variety of pain management techniques.

Looking back a hundred years to home births, babies were probably delivered in many different ways, with the mothers in many varied positions. Today in the developed world most babies are born in the hospital, since this has come to be considered safer. In the hospital it is easier to monitor the labor process and the cardiac activity of the fetus, as well as its precise position. Under these conditions the woman is more likely to spend her labor in bed, where she delivers in a semireclining position. In the last few years, even in hospital births, the trend has reversed toward allowing women a greater degree of freedom in the choice of labor position. Opinions remain divided, however, regarding the safest, least painful, and most natural kind of childbirth from the perspective of both the expectant mother and her baby.

A vaginal exam, which is always performed with the consent of the mother, can provide important information about how labor is progressing.

If labor is proceeding normally, medical staff will perform a vaginal exam to assess the mother's progress about every three hours, if she wants. If her contractions are unproductive, prolonging labor, she will often get an intravenous drip of oxytocin to help things along.

< The first stage of labor can feel like it is taking forever. Standing up and walking around between contractions may feel good and help make labor more productive. Eating and drinking is also a good idea. The energy will come in handy soon.

Each contraction temporarily halts the circulation of blood in the uterine wall, making the baby dependent upon the oxygen already in the placenta. During a normal delivery, this poses no risk to the baby. Babies have a fantastic ability to cope with the stresses of birth. The adrenal glands excrete large amounts of both adrenaline and noradrenaline into the circulatory system, keeping the heart rate up and making it easier for the heart to pump.

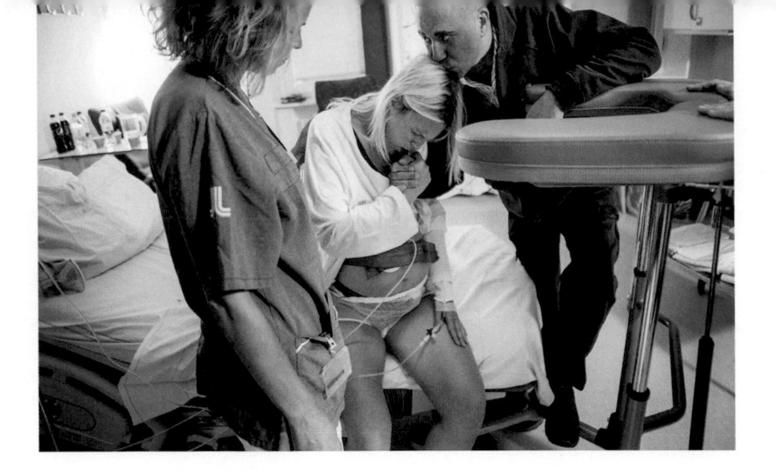

The contributions of the partner and doctor or midwife are very important at this stage. Their emotional and physical support is crucial.

This improves the blood, oxygen, and nutrient supply to the brain. Never again in a human being's life do so many stress hormones flow in the blood at once—it is a feat to be born!

The heart rate of the baby is usually measured by the external fetal monitor again during this stage of labor. Often it will be monitored right up to the moment of birth, using either an external ultrasound device or a scalp electrode attached to the baby's head. An ultrasound transducer will also be placed on the mother's abdomen to measure the frequency of her contractions. It is normal for the heart rate of the baby to slow at times, particularly late in labor as contractions occur, but normally the oxygen supply to the baby is quickly restored and the pulse picks right back up again.

If the heart rate pattern changes, some adjustments may need to be made. It may be a good idea for the mother to shift positions, perhaps moving to lie on her left side. If she has been receiving oxytocin to strengthen her contractions, the dosage may have to be decreased, and a uterine relaxant may be given instead to reduce stress on the baby. The doctor can see exactly how much oxygen the baby is getting by drawing a few drops of blood from the head: the lactic acid level is a good indicator of oxygenation.

Once the cervix is fully dilated, the second stage of labor begins. It is called the expulsion stage and is divided into two phases. The passive (or descent) phase begins when the cervix is fully dilated and ends when the head of the baby reaches the pelvic floor. This can take about two hours. This phase tests the patience of many mothers, but now the baby is almost here. When the active phase begins, the woman will begin to bear down until the baby is born. This phase can last up to an hour and usually goes faster for mothers who have already given birth.

It is often a good idea for the mother to change labor positions every thirty to forty minutes. She may find a walker helps when standing becomes difficult.

> When the contractions are just a few minutes apart, labor usually becomes quite painful.

(overleaf) Inhaling nitrous oxide is a method of pain management that allows the mother herself to determine the dosage.

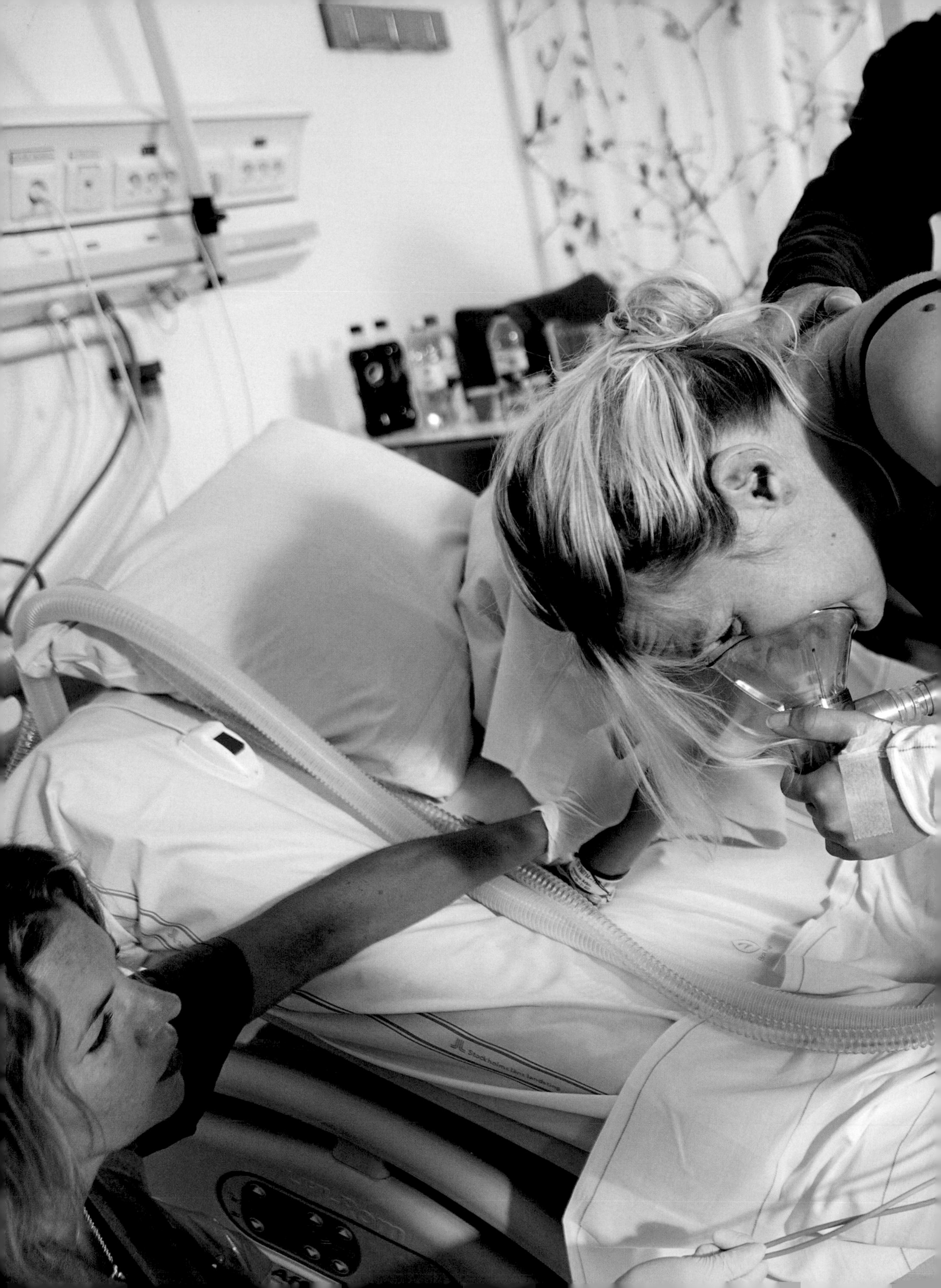
Stockholms

Labor pains

Historically, giving birth has been intimately associated with pain. This particular kind of pain, however, is different from almost any other. Moreover, most women forget their pain at the very moment their child is born. The result of the "labor pains," after all, is a perfectly unique baby of one's very own. Maternity care has long made available various kinds of pain relievers that are effective for the mother. Still, it is important to be aware that there is no such thing as a pain-free delivery, so preparing for some pain is a good idea.

Pain perception is highly personal, and those caring for the mother must be receptive to her needs. People always experience pain as more intense if they are frightened. It is thus very important that the birthing mother always be very well informed of what is happening at each stage. There is strong evidence that mothers who receive continuous support from a midwife or another caregiver such as a doula have a better experience during childbirth. They spend less time in labor and need less pain medication. They are also less likely to need drugs to stimulate their contractions, and more likely to give birth without aids such as a vacuum extractor, obstetrical forceps, or cesarean section.

Epidural anesthesia is the most effective method of relieving labor pains. It is always administered by an anesthesiologist.

Methods of pain relief

Confidence in the care provided

Fear and pain are intimately interrelated. A safe, friendly, expert environment is the most effective kind of pain relief. The presence and participation of the partner and the continuous support of a caregiver have been shown to reduce the need for medical pain relief.

Massage and heat treatment

These methods have been in use since time immemorial. Hand massage is excellent, and various ways of applying heat can relax and soothe the woman in labor. A hot bath or shower is another common method of pain relief.

Saline solution and TNS

Subcutaneous injection of small amounts of sterile water, or the administration of tiny shocks through the skin (transcutaneous electrical nerve stimulation, or TNS) appears to stimulate the endorphin system (the body's own painkillers) without negative side effects.

Acupuncture

Acupuncture (in which needles are inserted at specific pressure points) is another type of pain relief available at certain hospital maternity units. Recent years have seen less skepticism toward acupuncture, and many women are prepared to bear witness to its effectiveness.

Pain pills

Acetaminophen (the active ingredient in Tylenol) is very often used during the latent phase of early labor and also to ease afterbirth pains, caused by the uterus returning to its prepregnancy size. Aspirin increases the risk of bleeding and should not be taken. Stronger pain medication is available at the hospital if needed.

Morphine and other opioids

These types of pain medication act on the central nervous system and are not very effective against uterine pain. Also called narcotics, they have a generally sedative effect. They may be given as injections or in pill form. Their chief use is during the latent phase of early labor to allow the mother to relax and get some rest, but they may also be used to relieve pain after a cesarean.

Today these kinds of medications are used much less frequently during delivery. They are no longer thought to be particularly effective for pain relief, and they have negative side effects for both mother and child. Giving the mother morphine can affect the alertness of the baby and depress its cardiac activity and breathing. It also disrupts breastfeeding.

Nitrous oxide

A common and classic method of pain relief is the inhalation of a mixture of nitrous oxide and oxygen. The woman breathes through a mask she holds up to her own nose and mouth when she feels a contraction coming on. In the respites between the pains, she breathes normally, without the mask. In some women the intoxicating effect of this mixture may be anxiety producing.

Epidural anesthesia

Epidural anesthesia is a combination of powerful pain-relieving medication and an anesthetic that blocks pain impulses from reaching the brain.

About 97 percent of women find epidurals effective. They are most effective during the first stage of labor, before the cervix has fully dilated, and less effective once the bearing-down phase has begun. An epidural may be administered at any time during the dilation phase, up until the point when bearing down begins.

An epidural block is always administered by an anesthesiologist. Medical staff will monitor the woman's condition, including her blood pressure, heart rate, breathing, and ability to stand. Too much anesthesia can affect her blood pressure and breathing as well as her ability to move about. This could, in turn, have negative effects on the baby. Therefore, the baby too is continuously monitored using the fetal monitor.

Sometimes an epidural can prolong the expulsion stage of labor and make it necessary to deliver the baby by vacuum or forceps. The woman may also develop an itching from the medication, or have trouble urinating for a short time after the birth. Effects of the medication on the baby are thought to be minimal.

Pelvic anesthesia

Local anesthesia of the pelvic floor, known as a pudendal block, is used mainly when the delivery has to be completed with vacuum or forceps, or if an episiotomy or stitching of the perineum is necessary. Other forms of local anesthesia are also sometimes used. They may be administered as a spray or gel or as a direct injection.

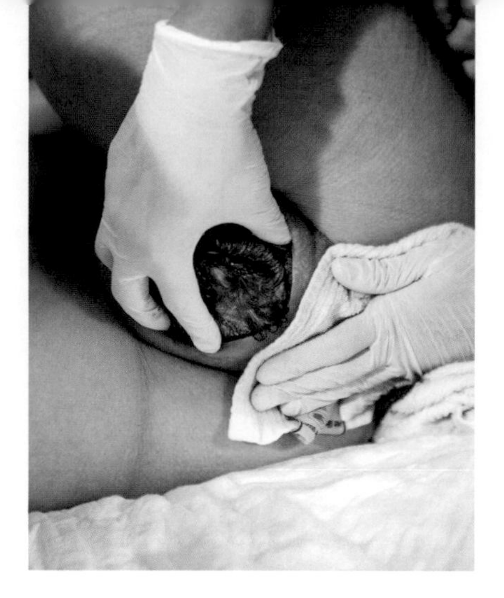

The head is visible in the vaginal opening.

Baby on the way!

In the active part of the expulsion stage, the mother feels the urge to bear down. It lasts from the moment the baby's head or bottom (whichever is coming out first) presses down on the pelvic floor until the baby is born. The mother often feels a sudden surge of pressure on her rectum, known as the bearing-down reflex. During this stage it is important for her to bear down actively when the reflex is activated during contractions, and to relax as much as possible between contractions, breathing deeply and calmly. Some women find it a relief to bear down, while others find this an extremely painful part of labor and are upset by the force of the reflex.

The head of the baby is delivered at the end of a contraction, followed by a pause. The shoulders come with the next contraction, perhaps assisted by the midwife or doctor. The rest of the body follows. Sometimes the baby will be born after just a few strong contractions, but often the expulsion stage is slower. It usually takes less than a couple of hours, although women giving birth for the first time may need more patience. At this point cooperation among all parties—mother, father, and doctor or midwife—is essential. The mother can reduce her risk of tearing by following the rhythm of her contractions and not trying to force the delivery. Staying calm and relaxed will help give her tissues time to stretch.

Sometimes a vacuum extraction technique is used to speed birth, using a suction cup made either of metal or of a rubberlike material. With the help of negative pressure from a pump, the suction cup is placed on the baby's head, and the physician or the midwife then slowly and methodically pulls, in time with the contractions and in the direction of the birth canal. Babies delivered this way often have a cup mark after birth, but vacuum is generally considered safe. An alternative aid is the obstetrical forceps.

(overleaf) Some women choose to give birth sitting, others kneeling. There are no rules, no absolutes, and no obvious choices. When it is time to bear down, the mother usually senses what will be best for her.

The doctor or midwife can use certain manipulations and other methods to reduce the risk of tearing during delivery.

> When the head is partly visible, the midwife can take hold of the baby's chin. A good grasp of the head will bring the rest of the baby out in just seconds. This picture also shows the scalp electrode used to monitor the baby's heart rate during delivery.

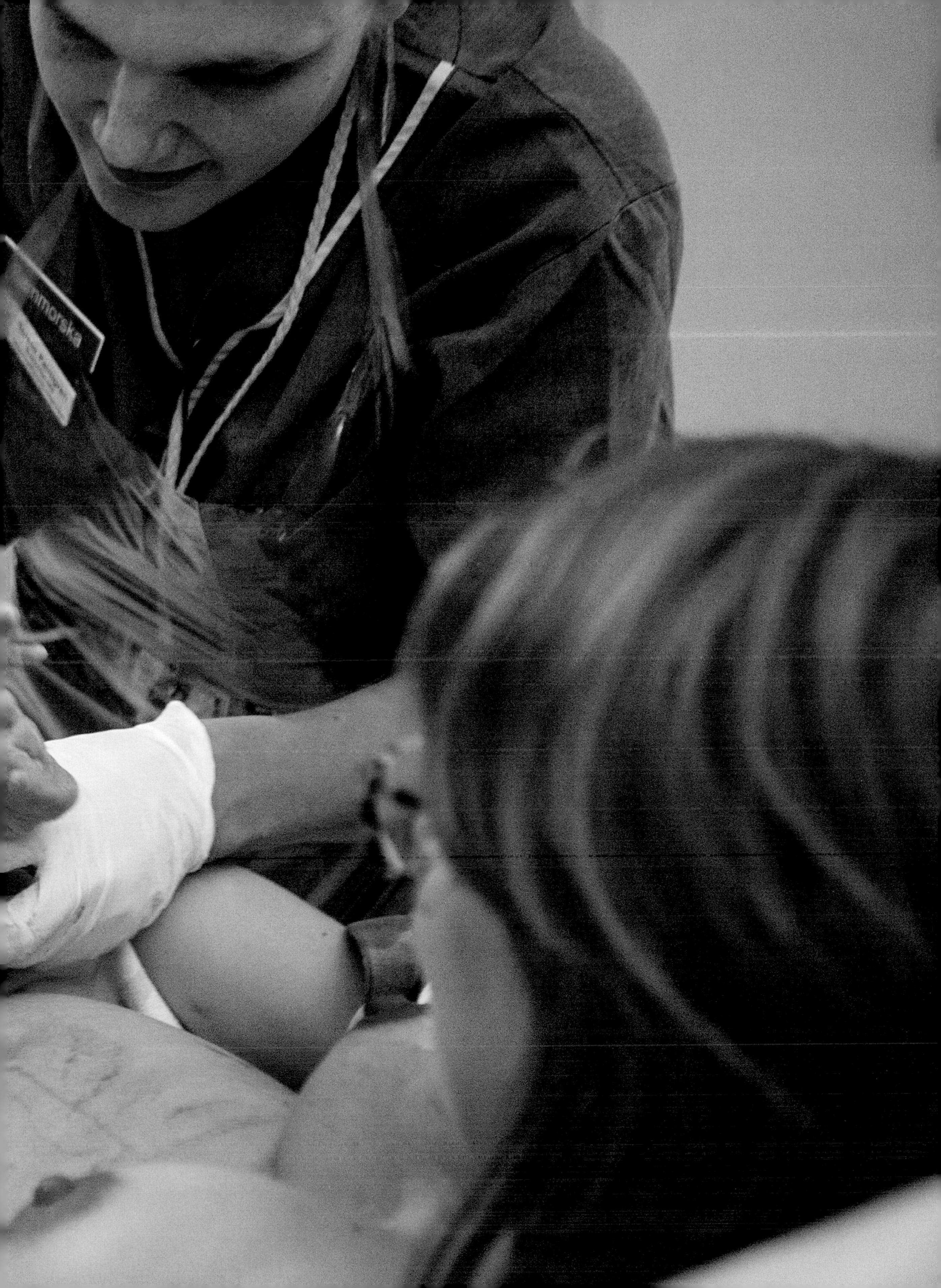

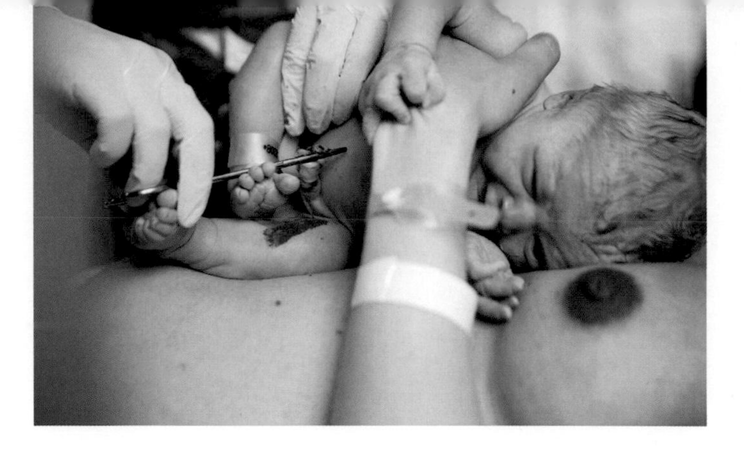

With one snip, nine months of intimate connection are severed.

Welcome to the world

The moment when the umbilical cord is cut and clamped is an enormous adjustment for the baby. The lungs now undergo their very first test. The baby's first cry draws air into the pulmonary cavities, and the crying and coughing reflexes release whatever mucus may be in there. Hormones also play a very important role in preparing the lungs for life outside the womb. Adrenaline reduces the accumulation of fluid that filled the cavities of the lungs in utero. One sign that the lungs are working properly is that the skin of the newborn becomes rosy, the muscles tense, and the cries get louder and harder. This is the moment when all the things the baby has been preparing for in the womb are put to the test.

Immediately after birth, medical staff will perform an Apgar test. This test scores the newborn's color, heart rate, grimace (or response to stimulation), muscle tone, and respiratory effort. The maximum score is 10, and most newborns receive a score of 9–10. The Apgar score will be carefully noted in the medical chart. A blood sample will also be taken from the umbilical cord to check the baby's oxygen levels.

Although the mother tends to be preoccupied by what is going on with her baby, her body still must complete the afterbirth stage: the discharge of the placenta and the fetal sac. This may take from a few minutes to an hour. An oxytocin injection shortly after the baby is delivered can help close the blood vessels in the placenta and prevent excessive bleeding. During this time the baby stays with the mother and has the opportunity to nurse, which helps the placenta exit more readily. The mother will feel afterbirth pains as the uterus returns to its prepregnancy size.

After delivery of the placenta, the mother is checked for any tearing or injury to the perineum or between her labia. Little tears will heal on their own, while more serious ones require stitches.

Today it is normal to wait a few minutes before cutting the umbilical cord to allow more blood from the placenta to reach the baby. The placental blood is rich in iron and carries important stem cells that will strengthen the baby's immune system later in life.

> Newborns routinely receive a vitamin K shot just after birth, subject to the mother's consent. Vitamin K helps the blood coagulate should any bleeding occur. The shot is given so early because babies are less sensitive to pain directly after birth.

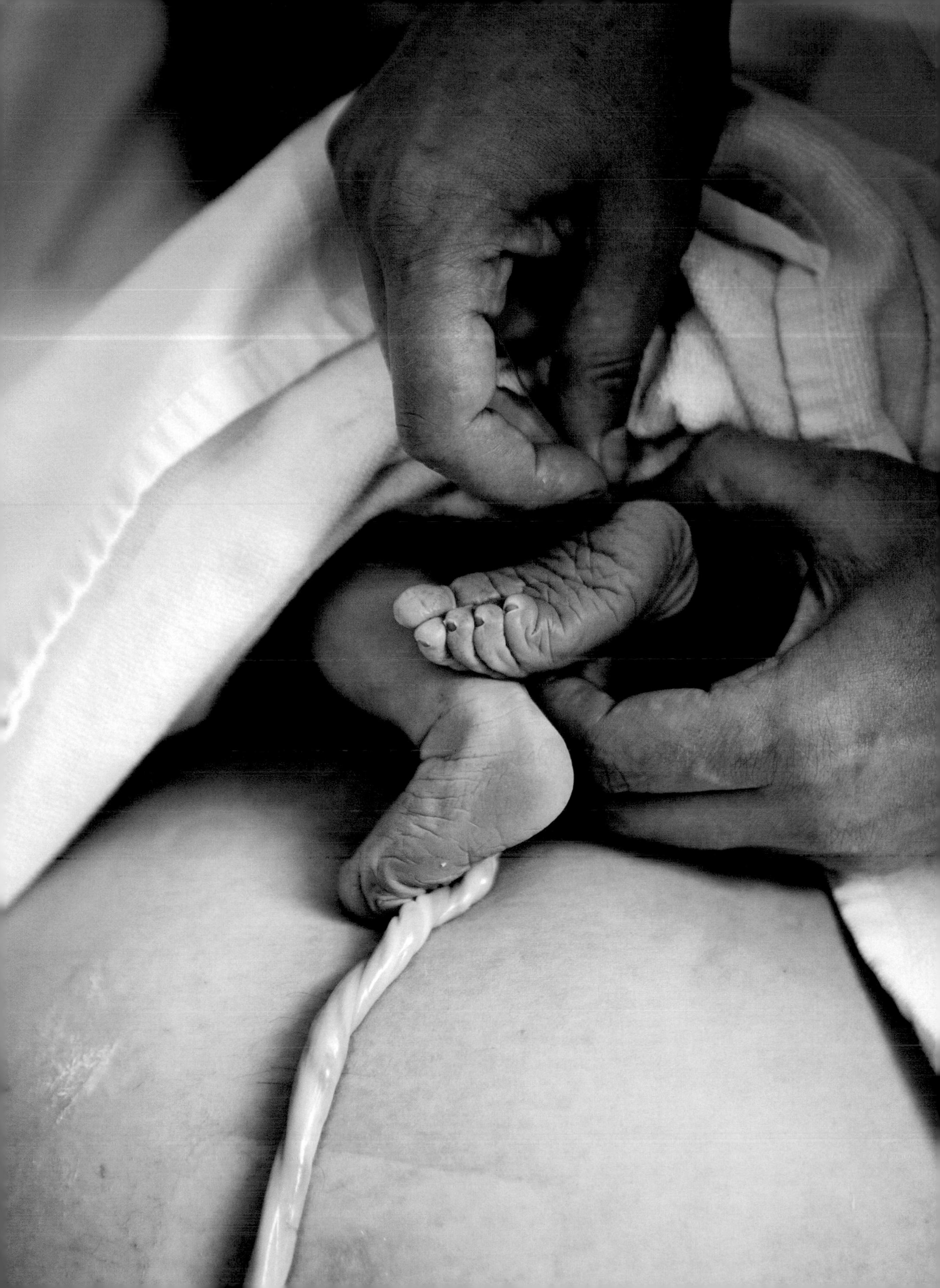

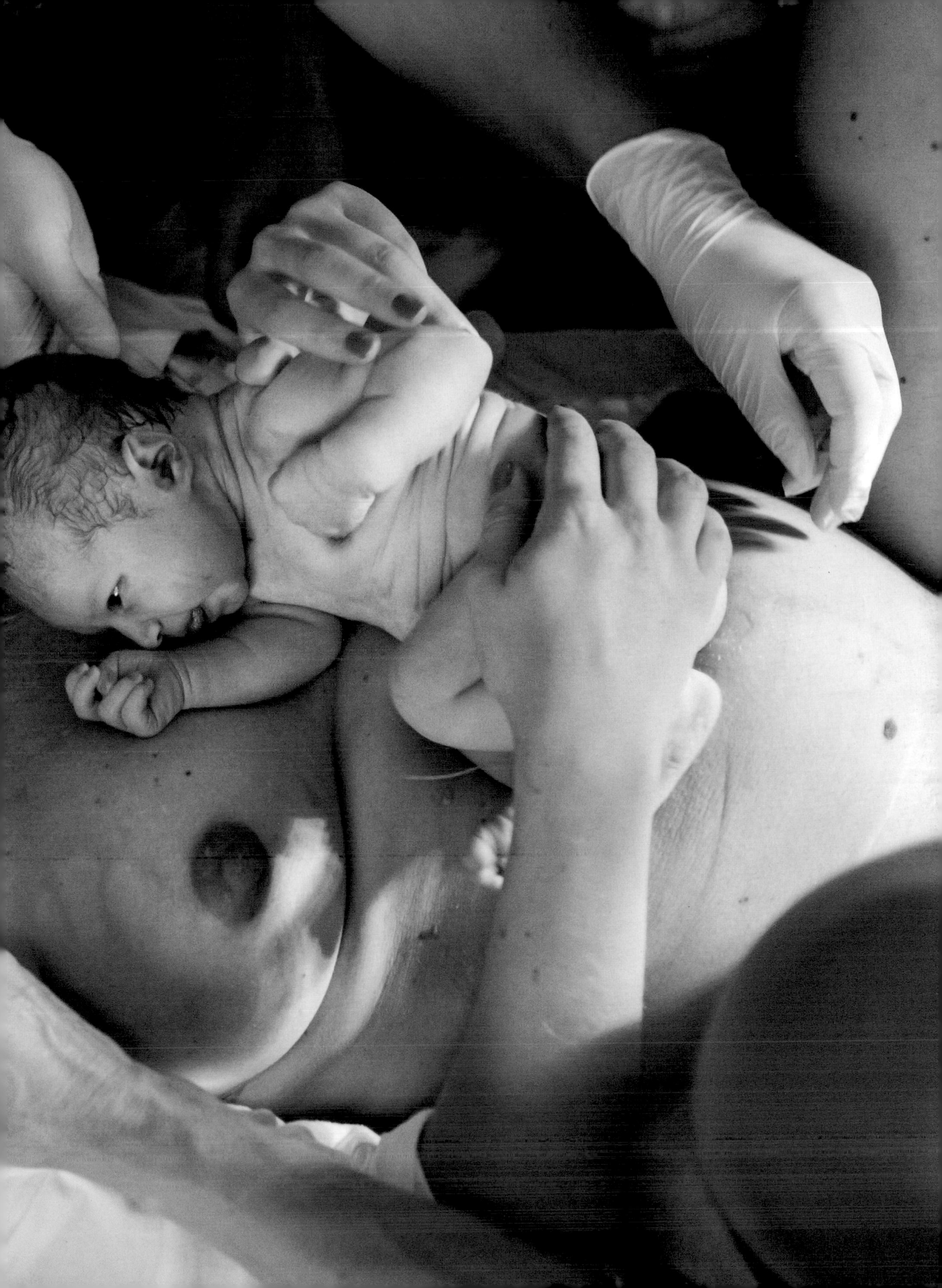

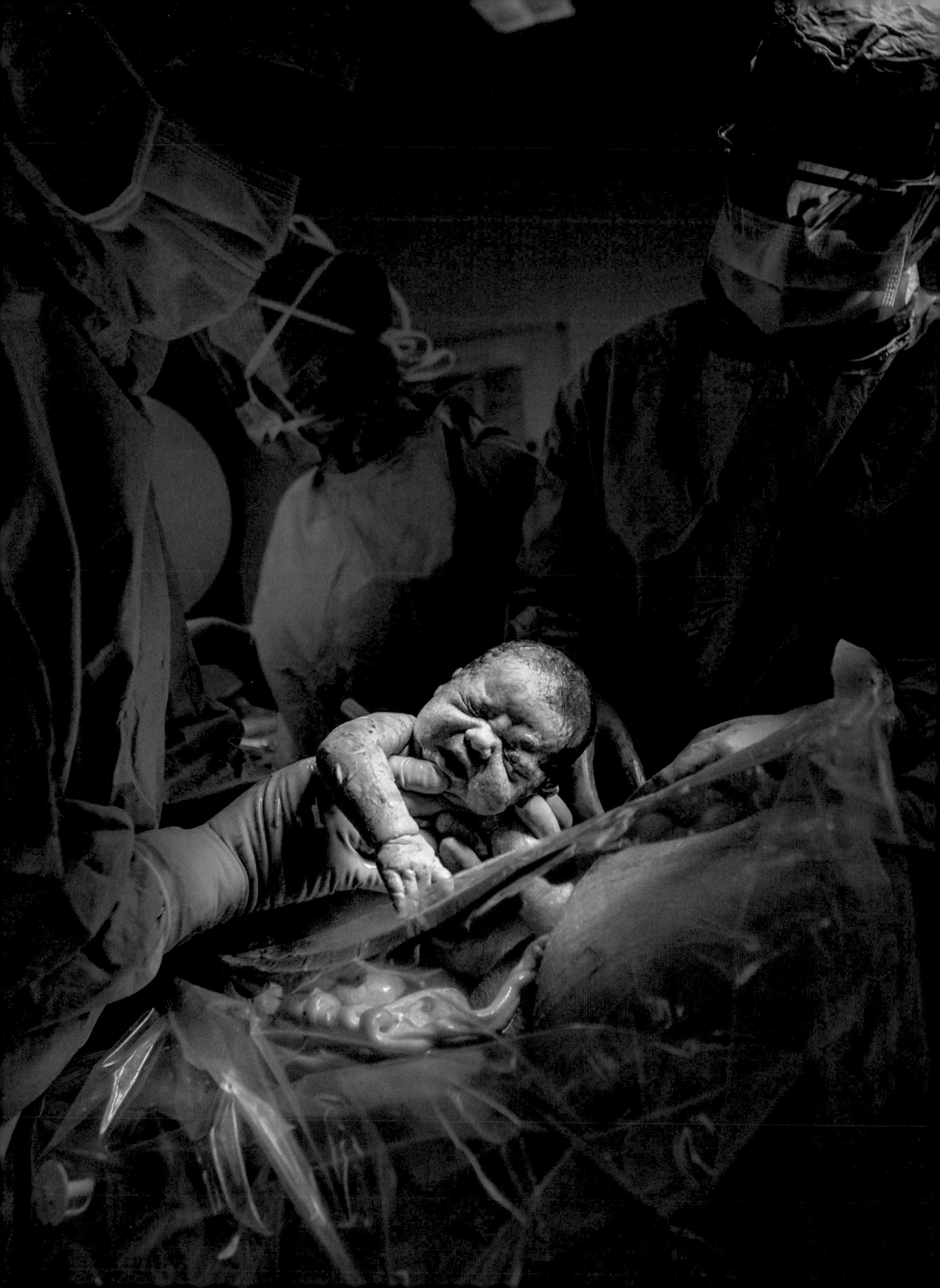

C-sectioning into life

In recent years the proportion of cesarean births has increased dramatically. In the United States it is now around 30 percent. There has even been a debate regarding whether it should be a woman's right to choose a cesarean. Unless it is an emergency, this decision is generally made by physician and mother/parents in consultation.

A cesarean may be planned well in advance or may become necessary during labor. Cesareans may be planned when the baby is too large in relation to the birth canal or is in the breech position, feet down, or, in very occasional cases, in transverse position. In some cases the baby may not be strong enough for a vaginal birth, or the mother may have a serious condition, such as preeclampsia, that makes a cesarean necessary. Fear of pain in childbirth has also contributed to the increasing number of cesareans performed today.

Unplanned cesareans are most commonly performed when the baby is experiencing distress (usually this means not getting enough oxygen in the womb), when labor is prolonged and the mother's contractions are very weak, when part of the placenta separates from the uterine wall too early, or when the umbilical cord has slipped into the vagina where the baby's head can compress it.

A cesarean delivery takes about 20 minutes from start to finish.

In emergency cesareans, when the health of mother or baby is at immediate risk, general anesthesia may be administered. In most other cases the mother is given spinal anesthesia, which eliminates all sensations of pain in the lower half of the body and also renders it immobile. For a cesarean the

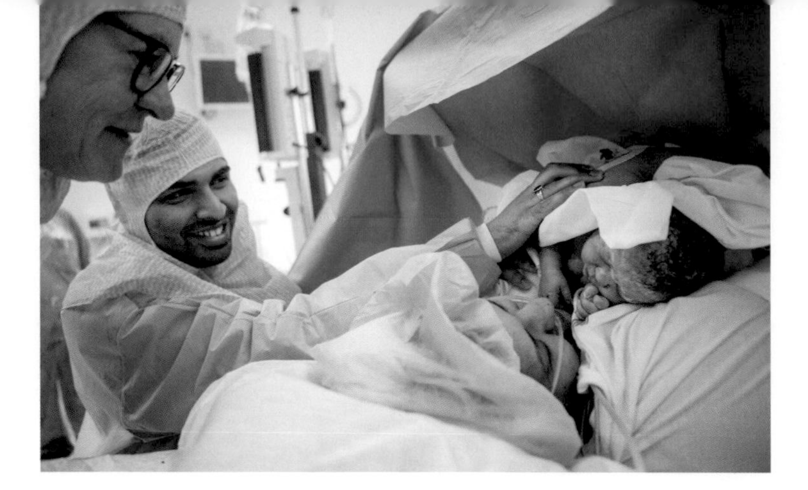

It is important for the baby to join the mother right away, even after a cesarean delivery.

surgeon usually makes a horizontal incision an inch or so above the pubic bone, opening the uterine wall and the fetal membranes so that the amniotic fluid rushes out. The child is carefully lifted out, and the umbilical cord is clamped and cut. Then the placenta is removed, and the uterus and abdominal wall are stitched. The mother remains fully conscious and can experience the moment of birth with her partner and hold her baby right away. If the baby is doing well, it may stay at her breast until the operation is over.

A mother who is delivered by C-section spends a few extra days in the hospital with her baby and partner. With modern surgical techniques, most women heal quickly if no complications arise.

After returning home, and for a few weeks, the mother will experience certain effects of the surgery and will need help. She will be more tired and less free to lift than a mother who has had a vaginal delivery.

Today most mothers can be awake during a C-section and can hold their newborns just a few minutes after the birth.

> The father or partner is almost always in the operating room to share the joyful moment when the baby arrives.

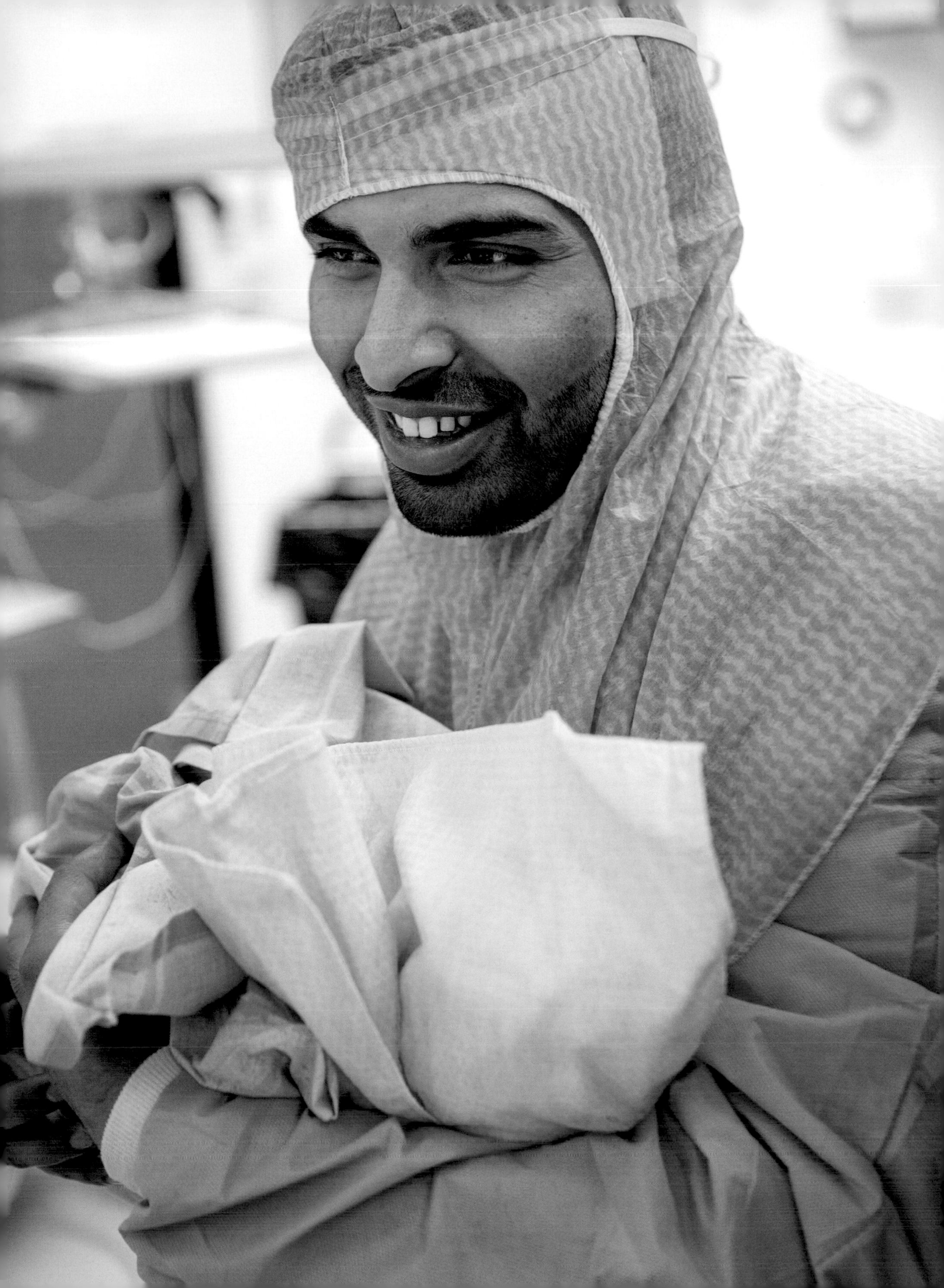

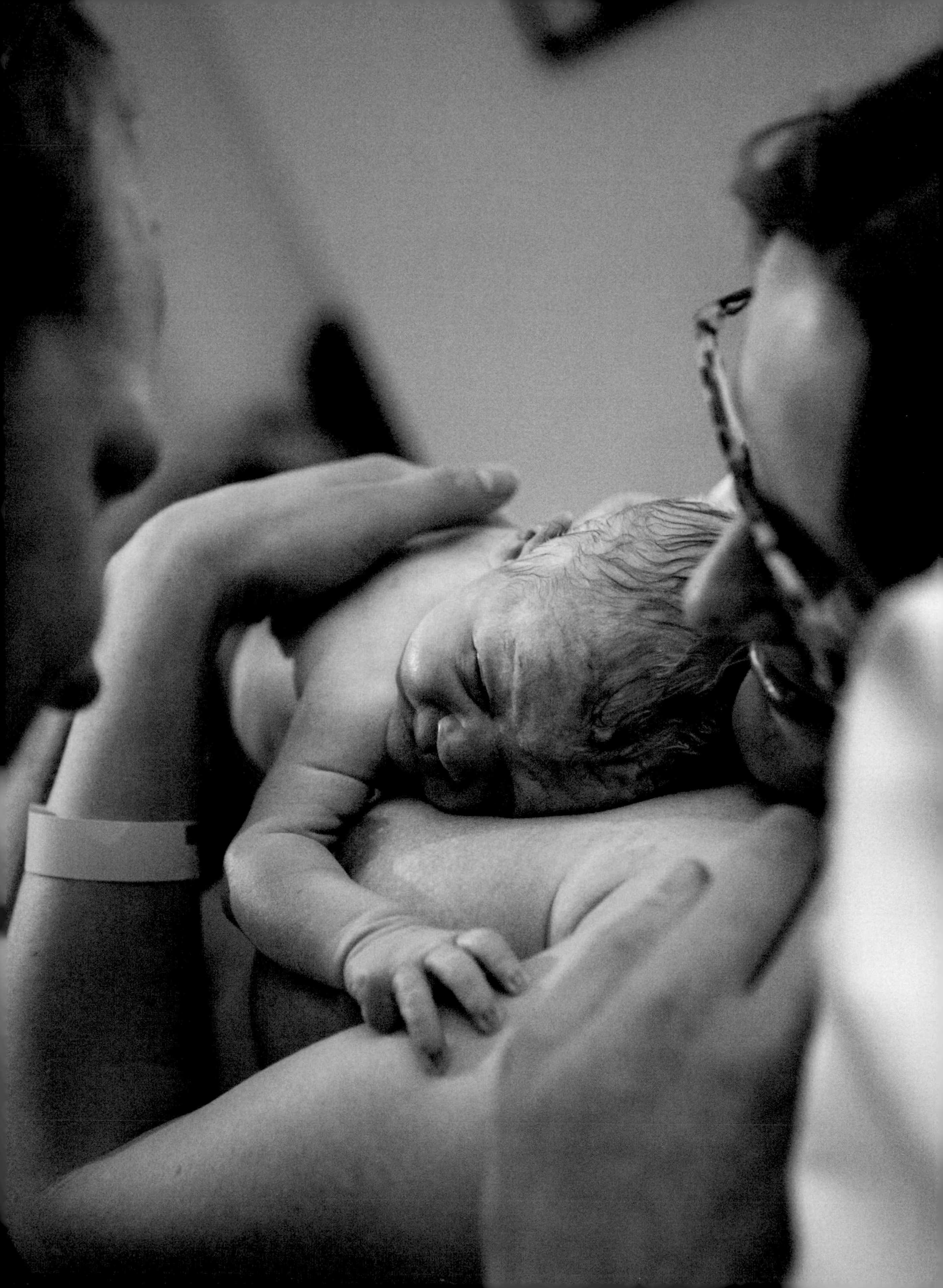

Aren't you lovely!

The delivery is over. On the mother's breast lies a little newborn, and the parents usually feel a great sense of liberation and relief. All their fears are past, and the mother's pains are over. Now it is time to take a deep breath and for baby and parents to begin, slowly and tenderly, to become acquainted.

Today hospitals are more sensitive than in the past to the importance of giving the new parents the peace and quiet they need with their baby immediately after delivery. Plenty of skin-to-skin contact is optimal for bonding and for getting breastfeeding off to the best possible start. As long as the baby is doing well, it is best if mother and baby stay together until the first feeding. Weighing and measuring the baby and doing other physical exams can wait; there is no hurry. More often than not, parents count the fingers and toes themselves.

Recent research has shown that newborns follow a sophisticated set of steps to their first meal. When placed on the mother's body, every newborn proceeds through nine instinctive stages:

The birth cry. It usually takes about 30 seconds for the baby to cry for the first time. Breathing independently is a huge adjustment.
Relaxation. The baby stops crying and the body relaxes. The eyes close.
Awakening. The eyes open and the baby makes small thrusts of the head.
Activity. The arms and legs start moving, the head lifts, and the baby's gaze fixes on the mother's breast.
Crawling. The baby makes crawling movements, often using the feet to push off.
Rest. The baby stops and takes a break.
Familiarization. If the areola is nearby, the baby will start licking to shape the nipple. Many newborns lick their own fingers too.
Suckling. The baby attaches to the nipple and sucks.
Sleep. After the baby finishes sucking, it is time for a long nap.

Putting the baby to the breast immediately after the delivery is very important for triggering milk production. For the parents this is also a wonderful moment, when they can marvel at their tiny, perfect newborn.

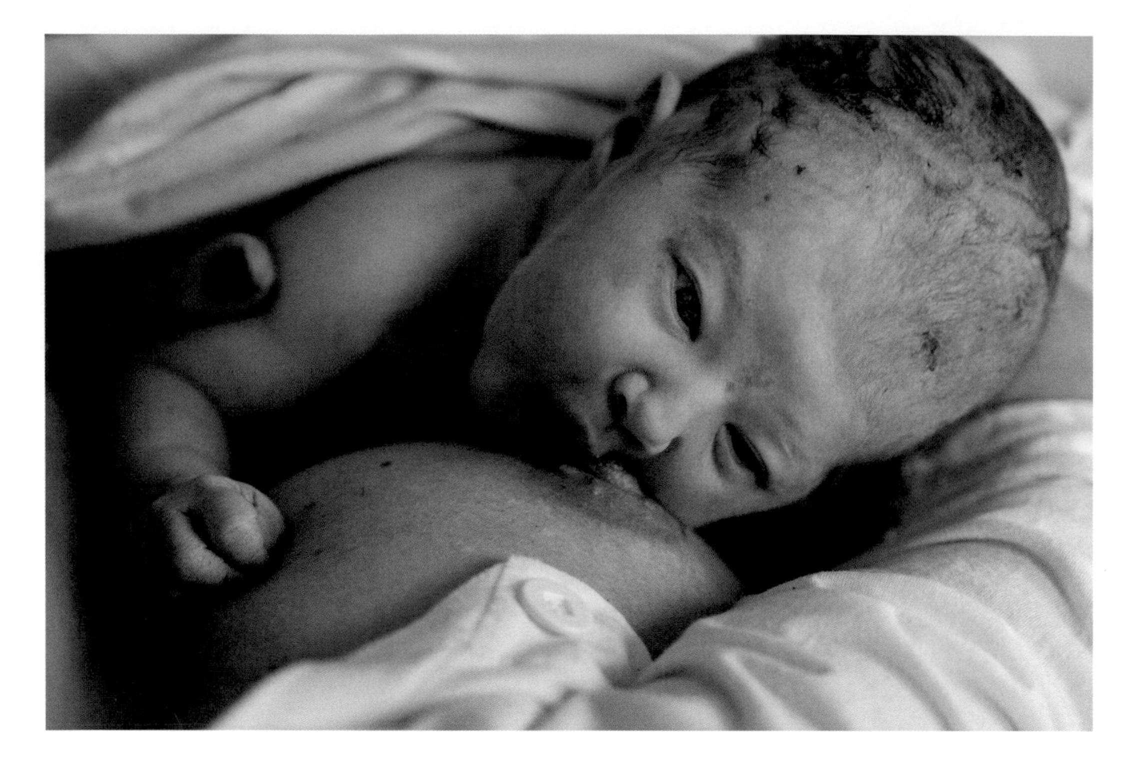

The baby instinctively searches out the nipple.

The baby's sucking is actually needed for the milk production and flow to begin. Nerve reflexes from the nipples are transmitted to special centers in the mother's lower brain, from which signals go to the pituitary gland, which begins to excrete prolactin, a hormone essential to milk production. Another hormone, oxytocin, affects the mammary glands, so that the milk is pressed out into the breasts. Oxytocin also helps the uterus to contract; the mother may notice contractions while breastfeeding, particularly soon after delivery.

Breastfeeding creates intimate contact between mother and child, and breast milk is the perfect food for babies, full of important nutrients and disease-fighting antibodies. Breast milk also contains substances that have a calming effect on the baby. When a mother nurses, oxytocin, a calming hormone, is released in her body, making her feel more harmonious as well.

The first milk is known as colostrum. Only a very small amount of colostrum is produced, but these drops are very precious to the baby, particularly in terms of early immune defense reactions. If the baby is reluctant to nurse right away, the mother can express this first milk by hand to help trigger her production. After two to three days the actual breast milk begins to flow. The earlier the child becomes used to taking the breast, the more quickly milk production will pick up.

The baby's gaze fixes on the nipple, and after some tentative exploration, the mouth latches on and begins to suck. Breast milk contains all the nutrients, vitamins, and minerals that the newborn needs. It also comes in a hygienic container at just the right temperature.

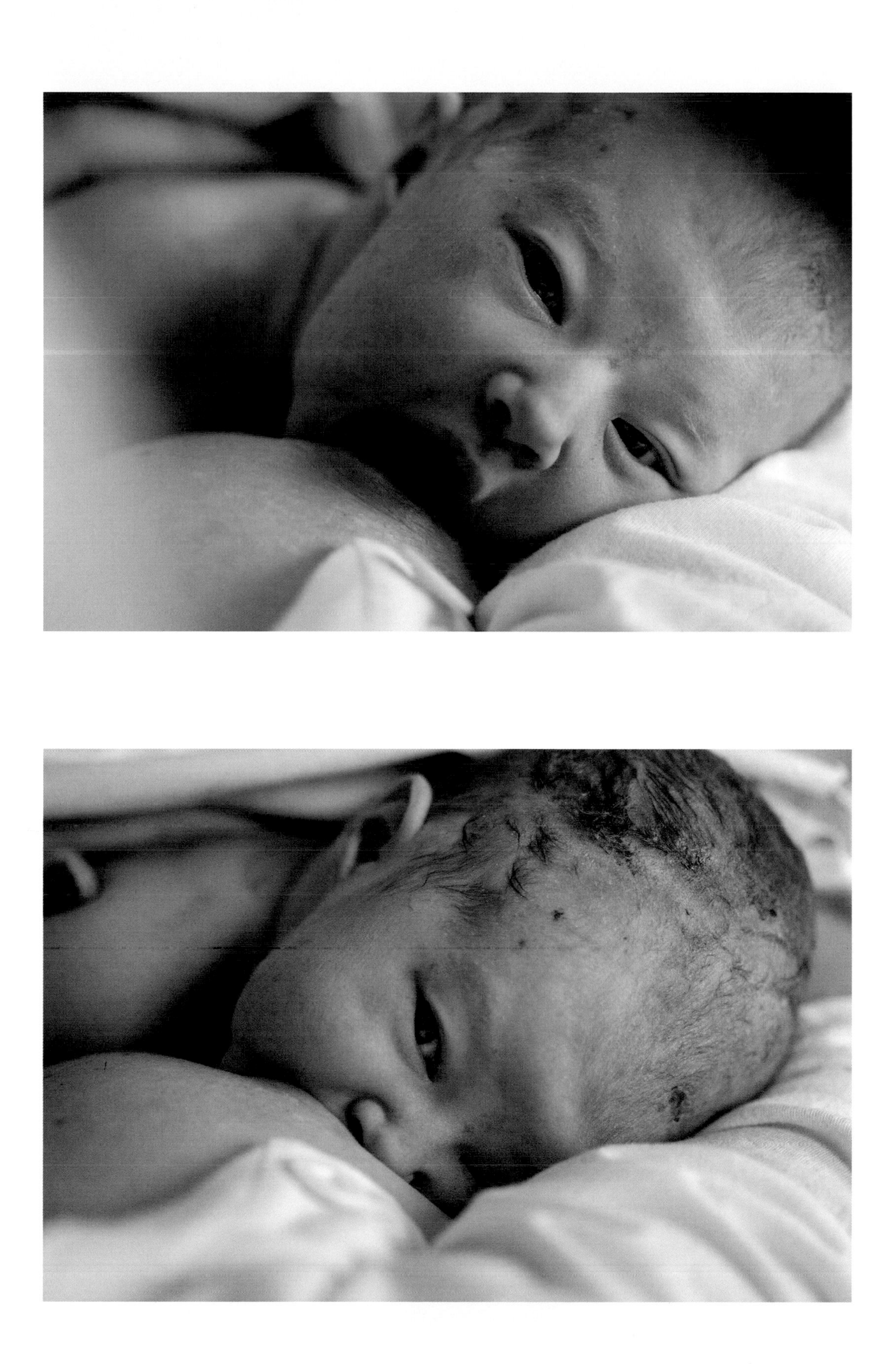

The first days

Before discharge a pediatrician examines the baby, taking a careful look at the whole body and listening to the heart and lungs. He or she also checks the muscle tone and reflexes—for example, the walking reflexes—because this reveals a great deal about the functioning of the baby's brain and nervous system. Other routine newborn screening tests include a pulse oximetry test, which measures how much oxygen is in the blood; a hearing test; and a blood test to screen for certain rare disorders, including phenylketonuria (PKU).

Most newborns have a slightly yellow tinge to their skin, owing to buildup of a chemical called bilirubin from blood cells that are no longer required. If the yellow tinge is more than moderate (infant jaundice), the baby receives light treatments (phototherapy) for a few days until the bilirubin is reduced.

Some families may leave the hospital after just six hours, but most will stay one or two days before going home. Arriving home for the first time with a new baby is a fantastic experience. The whole world looks different, feels almost unreal. Suddenly there is a new person constantly wanting love and attention. For the first few days everything focuses on the child, and time seems to stand still.

Caring for a newborn is both a great pleasure and a huge strain, particularly the first time around. Nursing and practical caretaking seem to fill every hour of the day and then some, and most first-time parents feel awkward and insecure. Why is the baby crying? Could I be doing something wrong?

Many mothers are often quite exhausted, so the help and support of the partner is particularly important. Nights are broken up with feedings until the baby begins to sleep through, which usually takes at least a couple of months. The baby then tends to sleep less during the day and shows more and more personality. He or she suddenly has habits and personal needs, as well as a will that may be strong indeed. It takes some time for the family to reestablish everyday routines and rhythms.

Six to eight weeks after delivery, the family makes a final visit to the obstetrician, to be sure that the mother has completely recovered from the physical aspects of childbearing. The obstetrician checks to see if the mother's pelvic floor has recovered from the birth, and they may discuss the right technique for pelvic floor exercises. The parents also have a chance to review the delivery, and to discuss their choice of birth control during the nursing period and afterward.

For the baby, visits to a pediatrician or a well-baby clinic are vitally important, more frequently in the early months, and once a year in late childhood. Such visits are an important source of support, and they help the parents solve any problems that might arise. Most pediatricians have regular call hours, and parents should keep emergency numbers for the doctor, hospital, and poison center available at all times. Parents' confidence in their own judgment and competence will show day by day.

New

Lennart Nilsson

A Child Is Born is the work that earned Lennart Nilsson (1922–2017) a place in history as a pioneer of medical photography. He began his career as a photojournalist in the 1940s. His first major photo essay to be published internationally was "A Midwife in Lapland" in 1945. Over the years that followed he honed his sense for visual storytelling.

Nilsson developed an interest in science and technology early in his career. He conceived the idea of a story on life in the womb in 1952, and pitched his vision to an editor at *Life* magazine the following year. Beginning in 1958, it took him seven years to finish the project, in collaboration with doctors, researchers, and other staff at five different Stockholm hospitals.

Nilsson sought out the technology he needed for his groundbreaking work. To reveal the life of the fetus he used macro lenses and optical microscopes, later turning to scanning electron microscopes and extremely wide-angle endoscopes. At his base at the Karolinska Institute in Stockholm, what began as an article in a magazine evolved into a life's work. Lennart Nilsson dedicated more than fifty years of his life to capturing the story of how we are born, as well as many other secrets of the innermost workings of our bodies. His images of the very first days of our lives are now considered among the most influential in the history of photography, and they have changed the way millions of people think about life before birth.

PHOTO: © AXEL ÖBERG

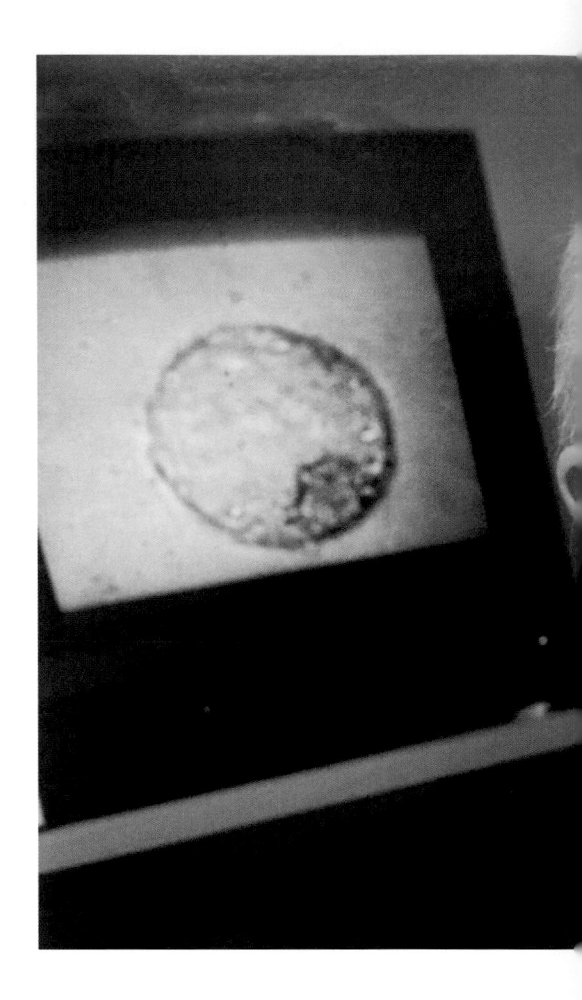

LINDA FORSELL is an award-winning freelance photojournalist from Stockholm. Between 2011 and 2015 she was based in New York and covered the North American continent for such major Scandinavian news outlets as *Dagen Nyheter* and *Dagens Naringsliv.* Her photos have been published in many international magazines and shown at exhibitions worldwide, including in China, Great Britain, and the United States.

LARS HAMBERGER is professor emeritus of obstetrics and gynecology at the University of Gothenburg. He is the author of the most recent editions of *A Child Is Born* and served as a medical expert and consultant to Lennart Nilsson on these books as well as other scientific book and film projects. Under Hamberger, a Gothenburg research team became the first in the Nordic countries to achieve in vitro fertilization, leading to the first IVF birth in Sweden, in 1982.

PHOTO: © LENNART NILSSON

PHOTO: © STEFAN TELL

Gudrun Abascal is a senior midwife with more than forty years' experience in maternity care. In her daily work she has helped and supported innumerable parents through pregnancy and childbirth. She has written books for expectant parents as well as healthcare professionals. She has founded maternity clinics that have helped to change and improve healthcare for expectant mothers. Her books are very popular and have been published in multiple editions.

Acknowledgments

The story of how we are born is one that unites us all as human beings. Perhaps it is for this reason that Lennart Nilsson found so many enthusiastic collaborators in his life's work.

Bonnier Fakta extends a warm thanks to the following people and organizations who assisted Lars Hamberger, Gudrun Abascal, and Linda Forsell in the preparation of this book: Anders Linde, Charlotte Hellsten, Erik Björck, Tina Råman, the Center for Fetal Medicine at Karolinska University Hospital, Ultraljudsbarnmorskorna, the obstetrics clinic and the reproductive medicine clinic at Karolinska Huddinge, the neonatal care division at Södersjukhuset, Yogashala, Mama Mia Karlavägen, and the Gullmarsplan midwife's office.

Anne Fjellström would like to thank Tommy Hallman and Anders Johansson for their help in processing Lennart Nilsson's images. Her thanks also to Alexandra Cornacchia Fjellström.

Finally, a special thanks to all those who agreed to be photographed during such a vulnerable and transformational time in their lives. Without them this book would not have been possible.

The U.S. publishers would like to thank Leila Schuler, MD, for consulting on obstetric practices in the United States.

Index

A
Aberration 76, 123, 126
Abdomen, growth 132, 159
Abortion 85
Acetaminophen 184
Acetylcholine 78
Acid indigestion 150
Acrosome 46, 51
Acupuncture 184
Adrenal glands 177
Adrenaline 78, 192
Afterbirth pains 184, 192, 202
Afterbirth stage 174, 192
Age, mother's, at pregnancy 22
Alcohol 100, 106, 108, 143
AMH (anti-Mullerian hormone) 65
Amniocentesis 120, 124, 127, 129
Amniotic fluid, the 83, 110, 118, 129, 134, 141, 143, 147, 156
Anabolic steroids 37
Apgar score 192
Apoptosis 16
Arm 89, 91
Arm bone 118, 120
Aspirin 184
Attraction 12
Axon 78

B
Backache 137, 150
Bath 177, 184
Bearing-down reflex 188
Bilirubin 206
Birth control pill 20
Birth weight 106, 144
Bladder 34
Blastocyst 58, 61, 68
  frozen 68
Bleeding 74, 173, 192
  light 74
Blood corpuscle 96, 100, 206
Blood count 114, 137
Blood flow analysis 127, 137
Blood pressure 114, 137, 159
Blood test 114, 192, 206
BMI (Body Mass Index) 134
Body development 76, 86, 92, 108, 120, 132, 159
Body temperature 27
  mother's 110, 132, 159
Bonding process 198, 201

Brain 76, 91, 92, 95, 108, 120, 141, 144, 178, 202, 206
  development 78, 83, 143
Brain cell 78
Braxton-Hicks contractions 159, 171
Breast 74, 104, 202
Breast milk 202
Breastfeeding 184, 201–202, 207
Breath, short of 137
Breech position 156, 197

C
Calcified cartilage 118, 120
Capacitated sperm 44
Capillaries of the placenta 96, 100
Cartilage 118–120
Cell division 16, 55
Cell nucleus 16, 20, 51, 52
Centriole 52, 55
Centromere 16
Cervix 23, 27, 44
  "ripeness" 165
Cesarean 114, 156, 162, 165, 182, 184, 197
  twins 162
Chance of pregnancy 27, 28
Chorionic villus sampling 124, 127
Chromosomal aberration 113, 126, 127
Chromosome 16, 19, 24, 52, 123
  in sperm 34, 37
Chromosome painting 19
Cilia in the Fallopian tube 28, 44, 55, 58
Circulatory system 76, 81, 96, 100
Cleft lip 86
Clitoris 117
Colostrum 202
Combined screening 124, 126
Confidence in the care provided 182, 184
Congenital illness 22
Contractions, weak 197
Corpora cavernosa 42
Corpus luteum 28, 58, 74
Cortisone 147
Crown rump length 92
Cytoplasm, egg 46, 50, 52

D
Dendrite 78
Depression 104
Diastasis symphysis pubis 150, 154

Dilated cervix 178
Dilation stage 174, 185
Dioxins 109
Discharge, vaginal 42, 137, 171
DNA 16, 22
  fetal (cell free DNA) 120
Dopamine 12, 78
Down syndrome 113, 123, 126
Drug to stimulate contraction 177, 178, 182, 185

E
Ear, development 95
Early pregnancy 22
Ectopic pregnancy 58
Egg culture 67, 68
Egg cytoplasm 46, 50, 52
Egg follicle 22, 24, 27, 28
Egg mitochondria 22
Egg shell 42, 46, 50, 51, 61
Emergency cesareans 197
Engaged head 156
Enzyme 28, 46
Epididymis 34, 37, 42
Epidural anesthesia 182, 185
Episiotomy 185
Estimated date of delivery 85, 114, 165
Estimation of fetal weight 127
Estrogen 12, 20, 27, 28, 42, 100
Exchange of nutrients and oxygen 96, 100
Exercise 134, 137
Expulsion stage 174, 178, 185, 188
Eye, development 95, 141, 143
Eyebrow 129
Eyelid 95, 118, 141, 143
Eyesight, fetus's 141

F
Face, development 86, 118, 143
Fallopian tube 23, 27, 28, 42, 44, 55, 58, 65
Fallopian tube funnel 23, 27, 28
Fear of labor pains 182, 184
Fetal membranes 192, 198
  stripping the membranes 165
Fetal monitor 173, 178, 185
Fetal sac 117, 129
Fertilized egg, *see* Egg and Blastocyst
Fibroid in the uterine lining 65
Filament of tubulin 52
Finger 89, 141

Fluid retention 134, 154
Follicle stimulating hormone 32
Food during pregnancy 104, 106
Foot, development 83, 86, 89, 120, 141
Fungal infection 137

G
Genetic abnormality 76, 126, 127
Genetic carrier testing 114
Genetic material 16, 20, 22, 32, 51, 52
Genetic test 126, 127
Genetics 15, 16
Genitals, development 117, 120
German measles (rubella) 109, 114
Germ layer 76
Growth, fetus 92, 108, 120, 132, 159

H
Hand 83, 86, 141
Hatching 61
hCG (human chorionic gonadotropin) 74, 85, 100, 126
Head of the sperm 32, 51, 52
Head position 156
Hearing, sense of 138, 143
Hearing test 206
Heart, development 76, 81, 83, 85
Heart muscle cell 81
Heartbeat (embryo's and fetus's) 81, 118, 159
  child's during the delivery 174, 177, 178, 188
Heartburn 150
Heat 184
Hemoglobin level 114
Hemorrhoids 154
Hepatitis 114
Hiccup 120, 143
Hormone injections (IVF) 65, 68
Hospital 206

I
ICSI (intracytoplasmic sperm injection) 68
Identical twins 15, 117
Illness during pregnancy 109
Immature egg 19, 20, 24, 120
Immature lungs 144, 147
Immature sperm 19, 34, 37, 120
Immune system 96, 192, 202
Implantation 61, 74
Incontinence 134
Induce labor 165, 173
Infant jaundice 206
Infection in the Fallopian tube 27, 65
Infertility 65
Inner cell mass 58, 76
Inner germ layer 76
Intercourse 27, 42, 173
Iron stores 114, 137, 192
Ischial spines 174
IVF (in vitro fertilization) 65–68

K
Kick, baby's 132, 141, 159
Kidney 76, 92

L
Labor pain 178
Labor-promoting hormones 173
Lamaze class 159
Lanugo 129
Leg, development 89, 120
Leg bone 118, 120
Leg cramps 150
Legs, tingly 159
Length measurement 92
Letter (birth) 174
Liver 83, 96
Lungs, development 76, 143, 144, 192
Luteinizing hormone 32

M
Massage 174, 177, 184
Maternity unit, hospital 162, 174
Medication 106, 108
Menopause 20
Menstrual cycle 20, 22, 24, 28, 74
Methods of pain relief 177, 184–185
Middle germ layer 76, 91
Midsection, sperm 32, 44, 51
Milk production 201, 202
Minerals 106
Miscarriage 22, 76, 102, 118, 123, 124
Mitochondria 22, 44
Morphine 184
Morula stage 58
Motion, fetus's 92, 120, 138
Mucous membrane
  in the Fallopian tube 55
  in the uterus 24, 28, 61, 74
Mucus
  in the cervix 27, 42
  in the vagina 42
Mucus plug 42, 171, 173
Muscle, fetus's 91
Muscle in the Fallopian tube 58

N
Narcotic drugs 108
Nausea 74, 104, 134
Neck, fluid behind the 126
Neonatal unit 144, 147
Nerve cell 76, 78, 91
Nerve impulse 91, 92
NIPT (noninvasive prenatal testing) 124, 126
Nitrous oxide 178, 185
Noradrenaline 12, 78
Nutrient cell 24, 27, 28, 42, 46, 55

O
Obstetrical forceps 182, 185, 188
Orgasm 42, 44
Outer germ layer 76
Ovaries 20, 22, 23, 24, 27, 28, 74, 76, 100
  fetus's 20, 120
  IVF treatment 65–68
  reserve 65
Overdue 165
Overweight 109
Oviduct 27, 28
Ovulation 20, 22, 27–28, 42
Oxygenation, fetus's 81, 96, 100, 147, 162
  at the delivery 177, 178, 201, 206
Oxytocin 12, 165, 173, 177, 192, 202

P
Pain 171, 174, 188
Pain pill 184
Pain relief, methods of 177, 184–185
Partner support 162, 184, 198, 207
PCBs 109
Pelvic anesthesia 185
Pelvic floor 134, 137, 207
Pelvic floor training 134
Penis 34, 42, 117
Period, *see* Menstrual cycle
Phenylketonuria (PKU) 206
Pheromones 12
Pinopode 74
Pituitary 24, 32, 74, 202
Placenta 81, 83, 96, 100, 118, 126, 127, 134, 147, 162, 165
  at the delivery 173, 177, 192, 197, 198
  barrier 106, 108
  hormone producer 85, 100, 173
  twins 117, 162
Planned cesarean 197
Polar body 24
Portio 23, 42
Position, delivery 177, 178, 188
Practice contraction 159, 171
Preeclampsia 114, 144, 197
Pregnancy signal 74
Pregnancy test 85
Premature baby 144–147
Premature birth 144–147
Prenatal appointment 113
Prenatal care 113, 159, 162, 207
Prenatal course 159
Prenatal diagnostic test 113, 124, 126
Prenatal exercise class 137
Progesterone 28, 58, 68, 74, 100, 173

Programmed cell death 16, 20
Prolactin 202
Proportions, fetus's 118
Prostaglandins 165, 173
Prostate gland 34, 42
Protein in urine 114, 137
Puberty 20–22, 32
Pudendal block 185
Pulse oximetry test 206

R
Reflex 138, 143
Removal of nutrient cells 46
Reproductive system 20–22, 32–37
Rubella (German measles) 109, 114

S
Saline solution 184
Scalp electrode 178, 188
Scrotum 117
Secretion from the prostate gland 42
Seminal fluid 34, 42
Seminal vesicle 34, 42
Seminiferous tubules 32, 34, 37
Sense of smell 141
Sex 19, 117, 120, 126
Sex cell 19, 32, *see also* Egg and Sperm
Sex chromosome 19, 117
Sex differentiation 91, 117
Sex hormone 12, *see also* Estrogen and Testosterone
Short of breath 137
Shower 177, 184
Signs of delivery starting 171
Skeleton 76
Skin 76, 129, 143
Skull bone 78, 92, 120
Sleeping pattern 92, 120, 143, 207
Sleeping position 159
Slot between shell and membrane 50, 51
Smells, sensitive to 74
Smoking 108
Sperm 19, 27, 28, 42, 44, 46, 50, 51, 56
Sperm production 34–37, 106
Sperm test 65
Spermatic duct 34
Spermatid 37
Spermatogonia 32
Spina bifida 123
Spinal anesthesia 197
Spinal cord 76, 91
Spine 83, 89, 91
Spotting 27, 74
SRY gene 117
Stage of labor 174
Stem cell 192
Stress 37, 109, 154
  baby's during delivery 178, 192, 197
Stress hormone 177, 178
Stripping the membranes 165
Sucking reflex 138
Sugar in urine 114, 137
Sugarlike molecule 61
Sweating 159
Swelling 134, 159
Symphysis-fundus measurment 132

T
Tailbone 89
Tail of sperm 32, 37, 44, 46, 51
Targeted ultrasound 126
Tearing 188, 192
Temperature, body 27
  mother's 110, 132, 159
Testicle 32, 34, 76, 120
Testosterone 12, 32, 120
Thumb-sucking 138
Thymus gland 32, 96
Tingly legs 159
Tiredness 104, 137
TNS (transcutaneous electrical nerve stimulation) 184
Toe 141
Toxins 37, 109
Tubule 44
Twin 117, 137, 156, 162
  identical (monozygotic) 15, 117
  nonidentical (dizygotic) 117
Tylenol 184

U
Ultrasound 85, 114, 120, 123, 124, 126
  twin pregnancy 117, 137
Umbilical cord 85, 96, 143, 147, 149
  at the delivery 192, 197, 198
Urine, fetus's 129
Urine test 85, 114, 159
Urethra 34, 42
Urinary tract 76
Uterine relaxant 178
Uterine tube 23, 28, 42, 44, 55, 58
Uterus 23, 42, 61, 118
  after delivery 192, 202
  space in 143, 149, 156

V
Vaccination 108
Vacuum extractor 182, 185, 188
Vagina 23, 42, 117
Vaginal discharge 137, 171
Vaginal examination 174, 177
Vaginal ultrasound 123
Varicose veins 154
Vas deferens 34
Vegan 106
Vegetarian 106
Vernix 129
Vitamin K shot 192
Vitamins 106

W
Walker 178
Walking and crawling reflex 143, 206
Water breaking 165, 171, 174
Weak contractions 197
Weight
  fetus's 102, 108, 143, 144, 156, 162
  mother's 104, 132, 134, 137
Well-baby clinic 207

X
X chromosome 19, 117
X-rays 108

Y
Yawn 120
Y chromosome 19, 117
Yoga 137
Yolk sac 91, 96, 110, 117

A Bantam Books Trade Paperback Original

Published in the United States by Bantam Books, an imprint of Random House, a division of Penguin Random House LLC, New York.

Bantam Books and the House colophon are registered trademarks of Penguin Random House LLC.

Fact checking: Anders Linde, Charlotte Hellsten
Graphic design: Patric Leo
Editorial: Per Wivall, Anne Fjellström, Susanna Eriksson Lundqvist
Prepress: JK Morris Production, Värnamo, Sweden

Original title *Ett barn blir till, Sixth Completely Revised Edition*
Bonnier Fakta, Stockholm, Sweden
Published in the English language by arrangement with Bonnier Rights, Stockholm, Sweden

ISBN 978-0-593-15796-1

Printed in China on acid-free paper

randomhousebooks.com

9 8 7 6 5 4 3 2 1